T0331155

Doping in Conjugated Polymers

Scrivener Publishing
100 Cummings Center, Suite 541J
Beverly, MA 01915-6106

Publishers at Scrivener
Martin Scrivener (martin@scrivenerpublishing.com)
Phillip Carmical (pcarmical@scrivenerpublishing.com)

Doping in Conjugated Polymers

Pradip Kar

Department of Applied Chemistry,
Birla Institute of Technology,
India

WILEY

Co-published by John Wiley & Sons, Inc. Hoboken, New Jersey, and Scrivener Publishing LLC, Salem, Massachusetts.
Published simultaneously in Canada.

For general information on our other products and services or for technical support, please contact our Customer Care Department within the United States at (800) 762-2974, outside the United States at (317) 572-3993 or fax (317) 572-4002.

Wiley also publishes its books in a variety of electronic formats. Some content that appears in print may not be available in electronic formats. For more information about Wiley products, visit our web site at www.wiley.com.

For more information about Scrivener products please visit www.scrivenerpublishing.com.

Cover design by Russell Richardson

Library of Congress Cataloging-in-Publication Data:

ISBN 978-1-118-57380-8

Printed in the United States of America

10 9 8 7 6 5 4 3 2 1

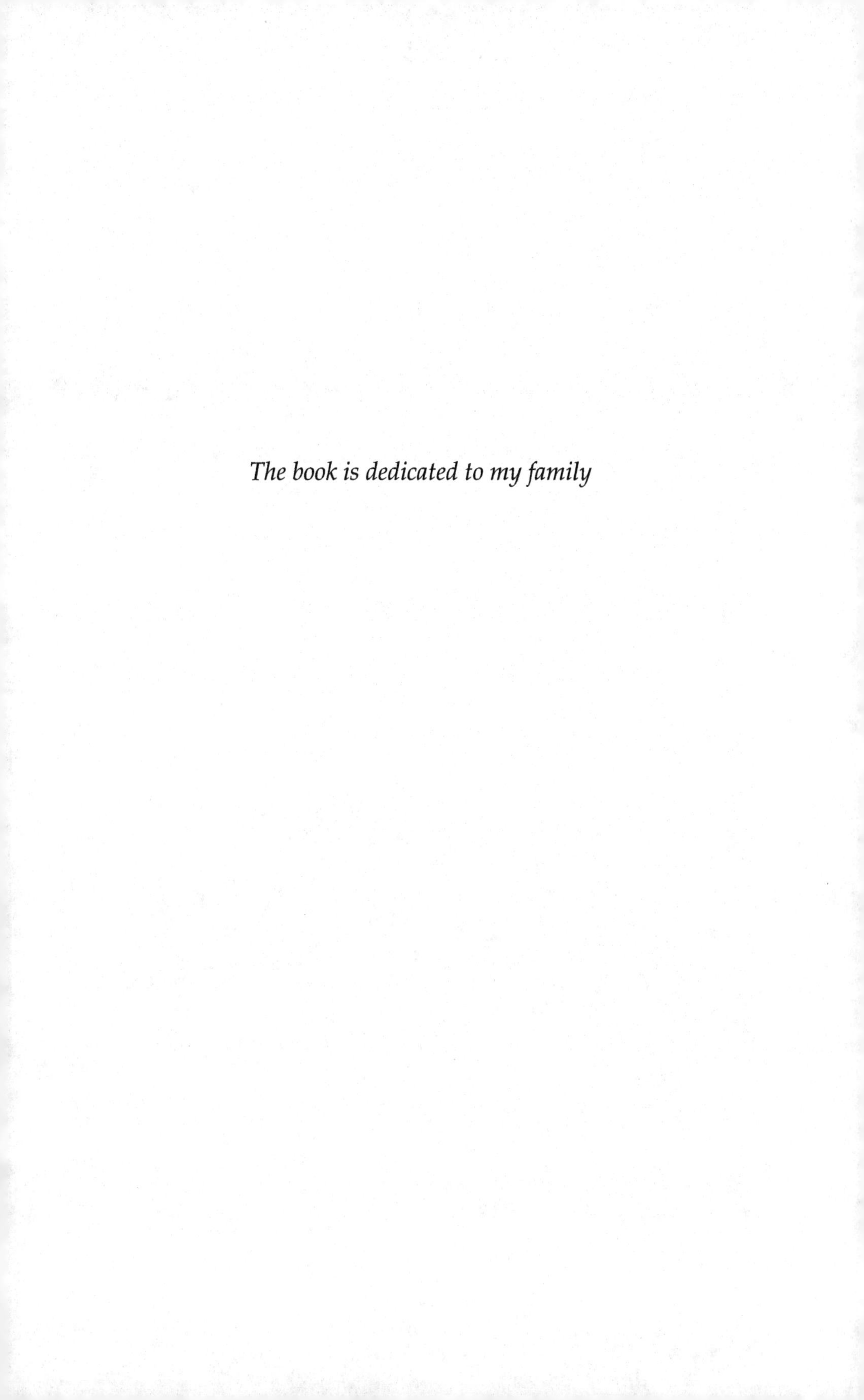

The book is dedicated to my family

Contents

Acknowledgement

I would like to express my sincere gratitude to Prof. Basudam Adhikari of the Materials Science Centre, Indian Institute of Technology, Kharagpur and Prof. Narayan C. Pradhan of the Chemical Engineering Department, Indian Institute of Technology, Kharagpur for their invaluable guidance, advice and constant inspiration. My acknowledgement is extended to Mr. Martin D. Scrivener and Scrivener Publishing LLC for this great opportunity.

Preface

The versatility of polymer materials has expanded as electroactive behavior has been included in the characteristics of some of the polymers. The most exciting development in this area is related to the discovery of intrinsically conductive polymers or conjugated polymers. Some examples are polyacetylene, polyaniline, polypyrrole, polythiophene, etc., as well as their various derivatives. The conjugated polymers which are a field of interest for researchers are also well known as "synmet" or "synthetic metal" due to the incorporation of some metallic characteristics, i.e., conductivity. Interest in this field is increasing day by day after the awarding of Nobel Prize for the discovery and development of electrically conducting conjugated polymers in the year 2000 by three scientists: Prof. Alan J. Heeger, Prof. Alan G. MacDiarmid and Prof. Hideki Shirakawa. Generally, the conductivity of these undoped conjugated polymers is 10^{-7}-10^{-11} S cm^{-1}. But for the application of conjugated polymers instead of inorganic or traditional semiconductors some higher conductivity is required. The conductivity of conjugated polymers, which are either weak semiconductors or insulators, increases by several folds due to "doping." These conjugated polymers convert to a conductor or semiconductor from the insulator or low semiconductor by doping. Although the conductivity of doped conjugated polymers is higher than that of saturated insulating polymers, it is much less than that of conducting metals, e.g., Cu, Ag, Au, etc., and most of the doped conjugated polymers show conductivity in the semiconducting region. However, it is universally agreed that the doping process is an effective method to produce conducting polymers. As doping makes a semiconducting polymer from an insulting or low conducting one, it is of very much importance for the real applications of the conjugated polymers as semiconducting material.

The performance of doped conjugated polymers is greatly influenced by the nature of dopants and their level of distribution within the polymer. Therefore, the electrochemical, mechanical, and optical properties of the doped conjugated polymers can be tailored by controlling the size and mobility of the dopants counter ions. The essential idea about the unusual nature of the species bearing charges, i.e., excited doped states of the conjugated systems, has been intensively discussed in the last twenty years. In this context the understanding of the nature of interaction by dopant with the π-conjugated systems is of foremost importance from the real application point of view. This rapid growth of interest in conjugated polymer-dopant interaction has been stimulated due to its fundamental importance to a cross-disciplinary section of investigators, chemists, electrochemists, biochemists, experimental and theoretical physicists, and electronic and electrical engineers. Finally, I wish to extend my sincere thanks and gratitude to all who helped me complete this project.

Pradip Kar

1

Introduction to Doping in Conjugated Polymer

1.1 Introduction

Recently, polymers have become the most widely used, versatile material on earth. This is due to some of the advantages they have over other materials such as flexibility, tailorability, processability, environmental stability, low cost, light weight, etc. [1]. Polymers are macromolecules which are formed by the repetitive union (mer unit or repeating unit) of a large number of reactive small molecules in a regular sequence. The simplest example is polyethylene, where ethylene moiety is the "mer or repeating unit" (Scheme 1.1). A major percentage of polymers are generally made up of carbon and hydrogen atoms with a minor percentage of some heteroatoms such as nitrogen, oxygen, sulfur, phosphorous, halogens, etc. In general, polymer is more than a million times bigger with respect to its size and molecular weight than that of small molecular compounds. The properties of polymers depend on their chemical composition, molecular structure, molecular weight, molecular weight distribution, molecular forces and

Ethylene Monomer | Polyethylene Polymer | Mer unit or Repeating unit

Scheme 1.1 Monomer and repeating unit for polyethylene.

morphology. Even in the fifth decade of the last century polymers were well known as electrically insulating materials. In modern civilization, polymers have been used as insulating cover on electrical wire, insulating gloves, insulating switches, insulating coatings on electronic circuit boards, low dielectric coatings, etc. [1]. The so called insulating polymers generally have a surface resistivity higher than 10^{12} ohm-cm. The polymers are insulating in nature due to the saturated covalent long-chain carbon framework structure or saturated covalent long-chain framework of carbon and some heteroatoms such as nitrogen, oxygen, sulfur, phosphorous, halogens, etc. In these polymers, the nonavailability of free electrons is responsible for their insulating behavior [2].

The versatility of polymer materials has expanded as electrochemical behavior has been included in the characteristics of some of the polymers. The electrochemical behavior means the mode of charge propagation, which is linked to the chemical structure of the polymer. In short, the chemical change within the polymer can help charge propagation, or the polymer can carry the charge through its chemical structure. The composites of conducting particles (carbon, graphite, metal, metal salt, etc.) with insulating polymers also show electrochemical behavior [3], e.g., composites of polyethylene oxide [4], polyethylene adipate [5] or polyethylene succinate [6] with Li salts. However, these materials are electrochemically active due to the electron transport within the conducting filler. In a true sense, the polymers themselves are not electrochemically active in these conducting composites. Based on the mode of charge propagation linked to the chemical structure of the polymer, the electrochemically active polymers can be classified into two main categories:

1. ion (proton)-conducting polymers, and
2. electron-conducting polymers.

In ion (proton)-conducting polymers the conduction of electricity is due to the transfer of proton which is present in its structure. Examples of ion-conducting polymers are sulfonated polystyrene, polyaryl sulfone, polyaryl ketone, etc. Similarly, the electron-conducting polymers can conduct electricity due to the transfer of electron through the polymer structure. The electron-conducting polymers can be further distinguished on the basis of mode of electron transport. In one type, the electrons can transport by an electron exchange reaction, i.e., reversible oxidation or reduction between neighboring electrostatically and spatially localized redox sites. Some examples are poly(tetracyanoquinodimethane) (PTCNQ), poly(viologens), some organometallic redox polymers, etc. The other type, which is the most exciting development in this area, is related to the discovery of intrinsically conductive polymers or conjugated polymer. Some examples are polyacetylene, polyaniline, polypyrrole, polythiophene, etc. [3] and their derivatives as well. The polymers having conjugated double bonds are intrinsically conducting, and these polymers can be oxidized or reduced using charge transfer agents (dopant) more easily for better electrochemical activity. The conjugated polymers which are a field of interest for researchers are also well known as "synmet" or "synthetic metal" due to the incorporation of some metallic characteristics, i.e., conductivity [7]. The interest in this field is increasing day by day after the awarding of the Nobel Prize for the discovery and development of electrically conducting conjugated polymers in the year 2000 by three scientists: Prof. Alan J. Heeger, Prof. Alan G. MacDiarmid and Prof. Hideki Shirakawa. This is because these conjugated conducting polymers have attracted attention of scientific and technical community for their use in various fields.

1.2 Molecular Orbital Structure of Conjugated Polymer

The molecular structure of the organic molecule is well explained in molecular orbital theory. In the organic molecule, the carbon 2s and all three of the 2p orbitals are combined first to make four new orbitals. This process of mixing is called hybridization and it generally also occurs in all the heteroatoms present within an organic molecule. The sp^3, sp^2, Sp hybridized carbon atom contains four single bonds, three single bonds with one double bond and two single bonds with two double bonds respectively. The single bond in the organic compound is constructed by a sigma bond, while the double bond is constructed by both a sigma (σ) and pi (π) bond. The σ-bond results from strong axial overlapping of hybridized electron cloud and π-bond results due to weak lateral overlapping of unhybridized p-orbital electron cloud. An example of such a type of simple organic molecule with double bond is ethylene. Here both the carbon atoms have three single bonds with one double bond. So, both the carbon atoms are Sp^2 hybridized and the double bond is formed due to lateral overlapping of perpendicular unhybridized p-orbital of both the carbon atoms. According to molecular orbital theory, the two π-electrons of the double bond in atomic orbital (AO) are placed into the bonding π-molecular orbital (MO), and anti-bonding π-MO remains vacant. The last filled bonding MO is known as highest occupied MO (HOMO) and the vacant lower anti-bonding MO is known as lowest unoccupied MO (LUMO). In the case of the simplest molecule with conjugated double bond, such as 1,3-butadiene, the picture is somewhat different. All the carbon atoms are also sp^2 hybridized here. In each atom one unhybridized p-orbital (generally p_z) remains perpendicular to the plane of carbon chain, and all those p-orbitals in each atom remain parallel to each other. So one p-orbital can laterally overlap to form π-orbital with either of the two nearest p-orbital, and ultimately the p-orbitals are delocalized (Figure 1.1). In light of molecular orbital theory, first the two bonding π-MOs and anti-bonding π-MOs in 1,3-butadiene are formed. Then, the two bonding π-MO with comparable energies and two anti-bonding π-MO with comparable energies are rearranged to two bonding and two anti-bonding π-MO due to delocalization (Figure 1.2). The HOMO and LUMO is now distinctly separated and the energy of HOMO is further enhanced, while the energy of LUMO is further decreased for conjugated 1,3-butadiene compared to that of nonconjugated form.

The molecular structure of conjugated polymer is comparable to that of conjugated organic molecule. A π-electron

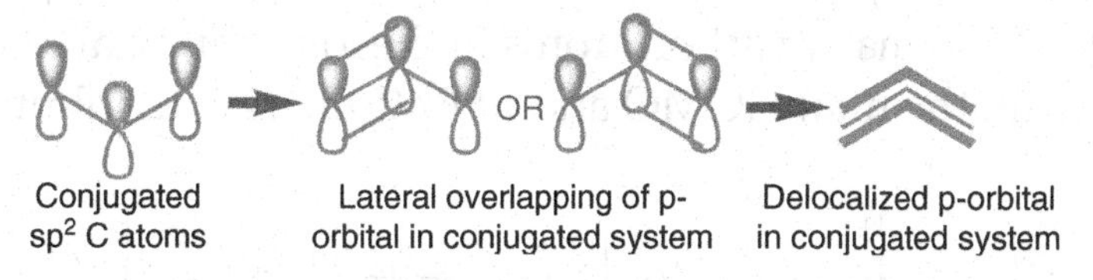

Figure 1.1 Delocalization of p-orbital in a conjugated carbon chain.

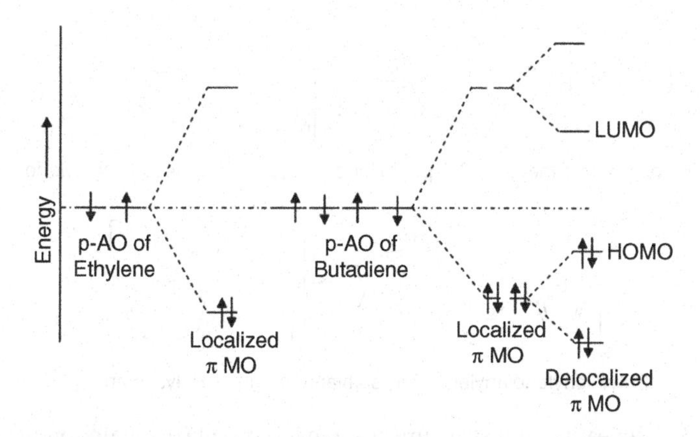

Figure 1.2 Comparison of π-MO energies for localized and delocalized systems in conjugated butadiene.

conjugated polymer contains alternative single and double bond throughout the chain. Some examples for such types of conjugated polymers are: polyacetylene, polyaniline, polypyrrole, polythiophene, polyfuran, poly(p-phenylene), polycarbazole, etc. The molecular structures of some of the conjugated polymers are shown in Scheme 1.2. Like other polymers, the conjugated polymer chain is also made of carbon atoms or carbon atoms with some heteroatoms such as nitrogen, oxygen, sulfur, etc., and all the atoms throughout the main polymer backbone are Sp^2 hybridized. Here, one p-orbital (unhybridized) in each atom also remains perpendicular to the plane of the polymer chain, and all those p-orbitals in each atom remain parallel to each other. So, ultimately the p-orbitals are delocalized throughout the polymer chain due to lateral overlapping of p-orbitals on either side. An example of such energy splitting resulting from the linking together of conjugated polythiophene is shown in Figure 1.3 [8]. Due to so many carbon atoms in the long conjugated polymer chain, the final HOMO and LUMO energy distribution is

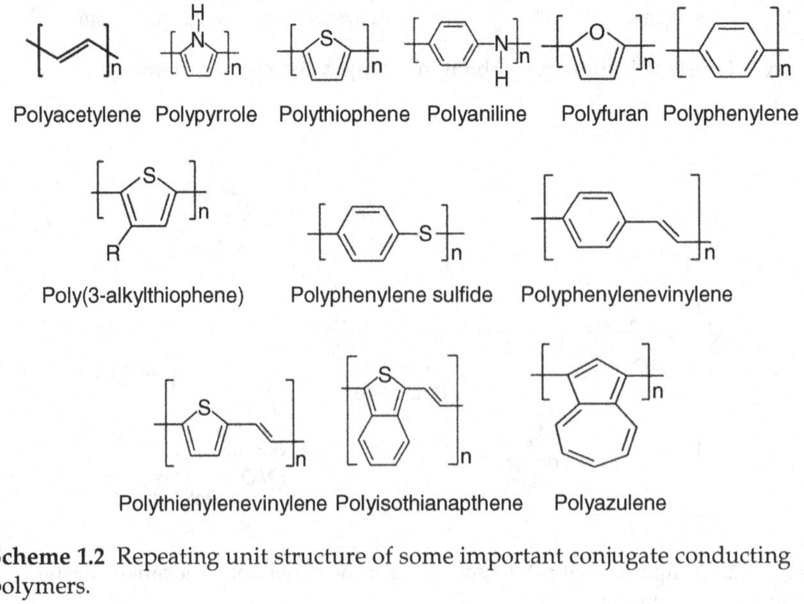

Polyacetylene Polypyrrole Polythiophene Polyaniline Polyfuran Polyphenylene

Poly(3-alkylthiophene) Polyphenylene sulfide Polyphenylenevinylene

Polythienylenevinylene Polyisothianapthene Polyazulene

Scheme 1.2 Repeating unit structure of some important conjugate conducting polymers.

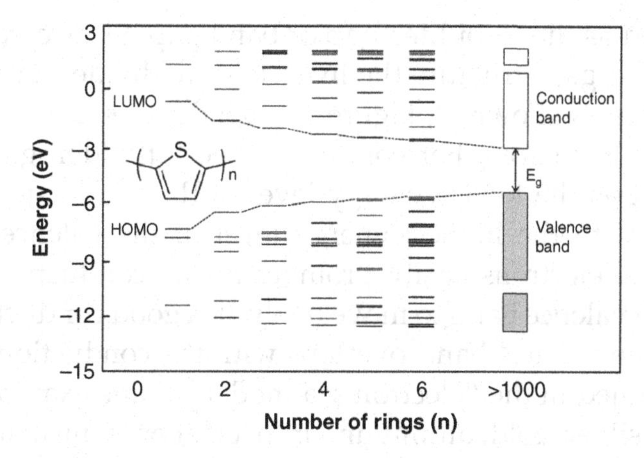

Figure 1.3 π-MO energy splitting due to delocalization in conjugated polythiophene. Reproduced with permission from ref. [8], Copyright © 1998 Elsevier Science Ltd.

very complicated. When the number of connected repeat units is very high, like in a long-chain conjugated polymer, splitting results in densely packed HOMO and LUMO energy states. The conjugated organic molecule shows different property due to this type of delocalization, which reduces the ground state energy of the molecule.

1.3 Possibility of Electronic Conduction in Conjugated Polymer

In general, electrons are responsible for the conduction of electricity by a material. More precisely, the electricity is carried by the free electrons within the material. The valence electrons, i.e., the electrons in the outermost shell of a material, can be placed in the lowest energy states designated as valence band. The electron in the conduction band is higher in energy than that of the valence band and it can move freely. For conduction of electricity by the electron, sufficient energy must be supplied to promote the electron to the conduction band from the valence band. The difference of energy between these two states, valence band and conduction band, is termed as the forbidden gap or band gap [9]. According to their band gap structure the materials are divided into three categories as shown in Figure 1.4: conductors, semiconductors and insulators. For conductors the forbidden gap does not exist as the lowest energy level of the conduction band merges with the highest energy level of the valence band. Thus, the electrons easily promote to the conduction band from the valence band to move freely. In a good conductor like metal, the valence band overlaps with the conduction band, as explained in the "electron sea model." Some examples are copper, silver, gold, aluminum, iron, etc. For an insulator like rubber, Bakelite, wood, saturated polymers, etc., the energy separation between the valence band and conduction band is large. Thus, the promotion of electron to the conduction band from the valence band is very difficult or not possible at all. However, the materials having moderate forbidden band gap have limited conductivity. These types of materials are known as semiconducting materials, e.g., GaAs, ZnO, conjugated polymers, etc. They have intermediate conductivity between the conductors and insulators.

For polymers the HOMO filled by electrons is denoted as valence band, the vacant LUMO is denoted as conduction

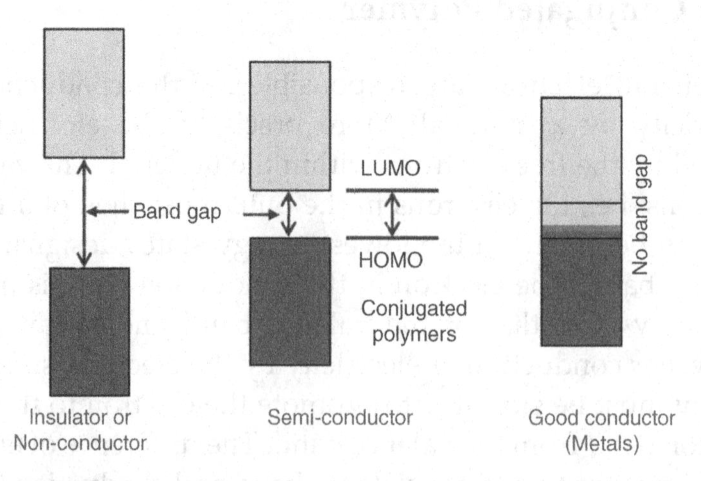

Figure 1.4 Band gap structures for insulator, semiconductor and conductor.

band and the energy difference between HOMO-LUMO is denoted as band gap [9]. The molecular orbital band gap for saturated insulated types of polymers is higher than 10 eV, which restricts the promotion of electrons from the valence band to conduction band in order for conduction to occur [2]. In the case of conjugated polymers the energy of delocalized HOMO is increased while that of the LUMO is decreased due to delocalization through the polymer chain. So, the band gap between HOMO and LUMO for the neutral stable (undoped) conjugated polymer becomes 3–6 eV and the conjugated polymers conduct electricity only in the lower semiconducting region. The conductivity of these undoped conjugated polymers is almost similar to that of insulator (10^{-7}–10^{-11} S cm^{-1}). But for the application of conjugated polymer as electrochemically active material, it requires conduction of electricity at least in the semiconductor region (10^{-2}–10^{-5} S cm^{-1}). Here, one of the important solutions is "doping," which reduces the band gap between HOMO and LUMO to 1–4 eV. So, in doped conjugated polymer the promotion of electron to the LUMO conduction band from the HOMO valence band is possible and hence now the conduction of electricity is also possible.

1.4 Necessity of Doping in Conjugated Polymer

The conjugated polymers cover a broad spectrum of applications from solid-state technology to biotechnology. Within the last few years, various electronic devices based on conducting polymers have been proposed. Possible applications of some of the important conducting polymers as electronic, optoelectronic, and semiconducting materials are listed in Table 1.1 [9, 10]. The most important criteria for selection of these materials are stability in ambient atmosphere, price, density, processability, etc. Conducting polymers having a very wide range of conductivity (10^{-5}-10^{3} S/cm) not only show high conductivity but also impart interesting optical and mechanical properties. Thus these types of conducting polymers may be used as a substitute for inorganic electronic, optoelectronic and semiconducting materials. The greatest advantages for the use of conducting polymers instead of inorganic materials are their easy synthesizability, architecture flexibility, tailorability, versatility, light weight, environmental stability, etc. However, in the neutral stable (undoped) state the conjugated polymers conduct electricity only in the lower semiconducting region and show very poor electrochemical activity. For the application of conjugated polymers instead of inorganic or traditional semiconductors, some higher conductivity is required than that of the undoped conjugated polymers. The conductivity of conjugated polymers, which are either weak semiconductors or insulators, increases by several fold due to

Table 1.1 Possible applications of some important conducting polymers [9].

Polymer	**Possible applications**
Polyacetylene	Rechargeable battery, photovoltaics, chemical and gas sensors, radiation detectors, Schottky diode, antielectrostatic, encapsulation, biotechnology, optoelectronics, solar cells.
Polypyrrole	Rechargeable battery, condenser, printed circuit boards, gas sensors, electroplating, Schottky diode, electroacoustic device, fillers, adhesive, transparent coating, electromagnetic shielding, elctrophotochemical cells, field-effect transistor, photocatalysts, physiological implantations, optoelectronics, conductive textiles.
Polythiophene	Rechargeable battery, display device, fillers, field-effect transistor, optoelectronics, Schottky diode, gas sensor, photocatalysts.
Polyaniline	Rechargeable battery, electrochromic devices, indicator devices, biosensors.
Poly (p-phenylene)	Rechargeable battery, fillers, photocatalysts.

"doping." These conjugated polymers convert to a conductor or semiconductor from the insulator or low semiconductor by doping. So, it can be said that;

$$\pi\text{- Conjugated polymer + Dopant} \longrightarrow \text{Conducting Polymer}$$

Although the conductivity of doped conjugated polymers is higher than that of saturated insulating polymers it is much less than that of conducting metals, e.g., Cu, Ag, Au, etc., and most of the doped conjugated polymers show conductivity in the semiconducting region (Figure 1.5). However, it is universally agreed upon that the doping process is an effective method to convert conjugated polymers to conducting polymers. As doping makes a semiconducting conjugated polymer from an insulting or low conducting one, it is of very much importance for real applications of the conjugated polymer as semiconducting material. The performances of doped conjugated polymers are greatly influenced by the nature of dopants and their level of distribution within the polymer. Therefore, the electrochemical, mechanical, and optical properties of the doped conjugated polymers can be tailored by controlling the size and mobility of the dopants counter ions. The essential idea about the unusual nature of the species bearing charges,

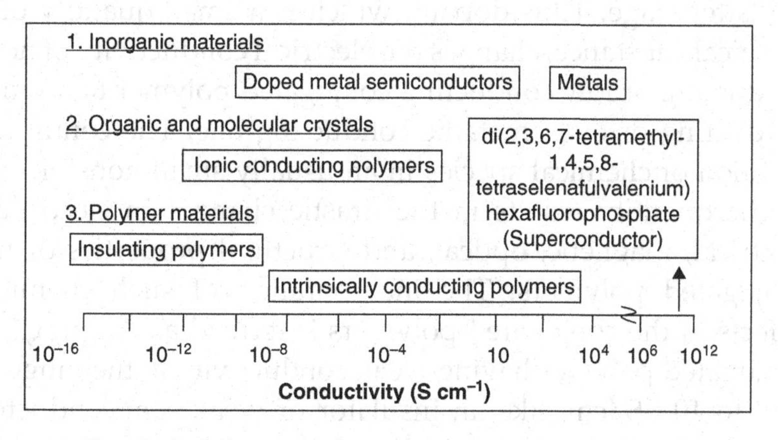

Figure 1.5 Electrical conductivity of materials.

i.e., excited doped states of the conjugated systems, has been intensively discussed in the last twenty years. In this context the understanding of the nature of interaction by dopant with the π-conjugated polymeric systems is of foremost importance from the real application point of view. This rapid growth of interest in conjugated polymer-dopant interaction has been stimulated because of its fundamental importance to a cross-disciplinary section of investigators, chemists, electrochemists, biochemists, experimental and theoretical physicists, and electronic and electrical engineers.

1.5 Concept of Doping in Conjugated Polymer

The first conjugated inorganic polymer, polythiazyl $(SN)_x$, was discovered in 1975 [11]. Polymer becomes superconductive after exposure in 0–29 K electrical field due to some charge generation within the polymer. However, the idea of using very low conducting conjugated polymer as an electrically semiconducting one actually emerged in 1977 with the findings of Shirakawa *et al.* [12]. They found that the conductivity of trans-polyacetylene was changed to 10^3 S/cm from 10^{-5} S/cm order after iodine vapor treatment. Here, the iodine incorporation into the polyacetylene is termed as "doping" in the latter stage. Thus dopant, which is a small quantity of a chemical substance, changes the electrical conductivity of non-conducting or low conducting conjugated polymer to a semi-conducting or even metallic conducting one. The controlled addition of chemical species in an usually small nonstoichiometric quantity results in the drastic changes in electronic, electrical, magnetic, optical, and structural properties of the conjugated polymers. The incorporation of such chemical defects in the conjugated polymers is termed as "doping." A conjugated polymer having weak conductivity in the range of 10^{-11} to 10^{-7} S/cm, like an insulator or weak semiconductor, is converted into a semiconductor or even a conductor-like metal (conductivity range 10^{-3} to 10^{3} S/cm) due to doping (Figure 1.5). The term doping is employed differently from that in inorganic or traditional semiconductors. Doping in inorganic semiconductors is generally due to the incorporation of holes in the valence band or addition of electron in the conduction band. Then the three-dimensional crystal lattice of the inorganic semiconductors is only slightly distorted due to doping in ppm level. However, the doping in conjugated polymeric semiconductors is somewhat different than that of inorganic or traditional ones. Generally in the conjugated polymer the counter ions are simultaneously inserted into the conjugated polymer matrix. Thus the doped polymers are considered as polymeric organic salts. The oxidizing or reducing agents, which convert polymer to polymer salt, are known as "doping agents" or "dopants." So doping in conjugated polymer means the addition of donor or acceptor molecule to the polymer. The name doping has caused some controversy in research for the following three reasons [13]:

1. The "dopant" counter ion with conjugated polymer neutralizes the extra charge created through electron removal or addition in the conjugated polymer by the doping process. However, in inorganic semiconductors doping is the incorporation of a little extra charge.
2. The ppm concentration of dopants is sufficient to dope an inorganic material, while the optimum amount of dopant used in conjugated polymers is within the general range of 0.1–0.5 mol %.
3. Generally there is little variety in the dopants available for inorganic semiconductors, while a wide variety of chemical species are used as dopants in conjugated polymers.

So, doping in conjugated polymeric semiconductors is somewhat different than that of inorganic or traditional ones.

However, for conjugated polymer the word doping is still well known, probably because of following three points:

1. It demonstrates the similarities like increase of conductivity due to the cause (doping), for conjugated polymers as well as inorganic semiconductors.
2. Doping changes the oxidation state of inorganic semiconductor as well as conjugated polymer with or without slight structural change.
3. The diffusion of charge carrier is the main mechanism for the conduction in both cases, and the higher dopant concentrations are used in conjugated polymers for the mobility of the charge carriers like doped inorganic semiconductors.

The above discussion refers to primary doping for a conjugated polymer which is phenomenologically different from the other two concepts of secondary doping and co-doping.

1.5.1 Concept of Secondary Doping in Doped Conjugated Polymer

A secondary dopant is an apparently "inert" substance which further increases the conductivity when applied to a primary-doped conjugated polymer. The substrate is inert in the sense that it is not directly interacting with the conjugated polymer-like dopant. The concept is illustrated for inorganic acid-doped polyaniline and its derivatives with phenolic derivative as secondary dopant [14]. In the secondary doping process, the "compact coil" of primarily doped conjugated polymer has been changed to "expanded coil" due to molecular interactions between doped conjugated polymer and secondary dopant. The expanded coil in doped conjugated polymer is considered as free carrier tail, which arises from delocalization of electrons in the polaron band. This free carrier tail is attributed to an increase in conductivity up to several orders of magnitude. The evidence of such a type of expanded coil during secondary doping is reported from spectroscopic and viscosity measurements. As for example, the dodecyl benzene sulfonic acid-doped polyaniline composite with novolac shows a red shift with the increase of novolac composition, indicating the secondary doping effect from phenolic polymer [15]. From the above discussion it is clear that the secondary dopant is an "inert" substance, which only enhances the performance of the primarily doped conjugated polymer. However, without using any primary doping, this type of secondary doping is not possible. That means camphor sulfonic acid-doped polyaniline thin films responded to secondary doping by o-chlorophenol, whereas without camphor sulfonic acid doping the o-cholophenol did not have any influence on the conducting property of the polymer [16]. Thus, this phenomenon can only be explained as an effect of so called secondary dopant material on the primarily doped conjugated polymer.

1.5.2 Concept of Co-doping in Conjugated Polymer

Co-doping is the process where both dopants are used to dope the conjugated polymer simultaneously. The co-doping process is required to improve the property and performance of the conjugated polymer. As for example, electrodeposition of polyaniline was made in different acid media (H_2SO_4 and $HClO_4$) in the absence and presence of two organic acid dopants, disodium salts of naphthalene-1,5-disulphonic acid and catechol-3,5-disulphonic acid. In that study the effect of single dopant and that of organic acid dopant with inorganic acid dopant on the properties of polyaniline has been studied. The author has explained this phenomenon of influencing the properties of primary doped conjugated polymer in the presence of a second dopant as "co-doping" [17]. Relatively large size dopants such as dodecyl benzene sulfonic acid or naphthalene sulfonic acid reduce the interaction between conjugated polymer chains, giving rise to soluble in various organic solvents. However, mechanical properties and conductivity of the soluble conjugated polymer doped with large size dopant are relatively poor. On the other hand, the conjugated polymer doped with smaller counter ion-containing dopants like HCl, H_2SO_4, $FeCl_3$, etc., shows higher conductivity but poor solubility, stability and mechanical property. Co-doping in conjugated polymer is helpful in combining properties like conductivity, mechanical strength, solubility, stability, etc., thus improving the overall performance of the doped conjugated polymer during application. Improved ammonia-sensing property is reported [18] for polyaniline thin film co-doped with HCl and p-toluene sulfonic acid over the polyaniline doped with only p-toluene sulfonic acid or only HCl (Scheme 1.3). Therefore, co-doping in conjugated polymer can be described as the combined effect of two

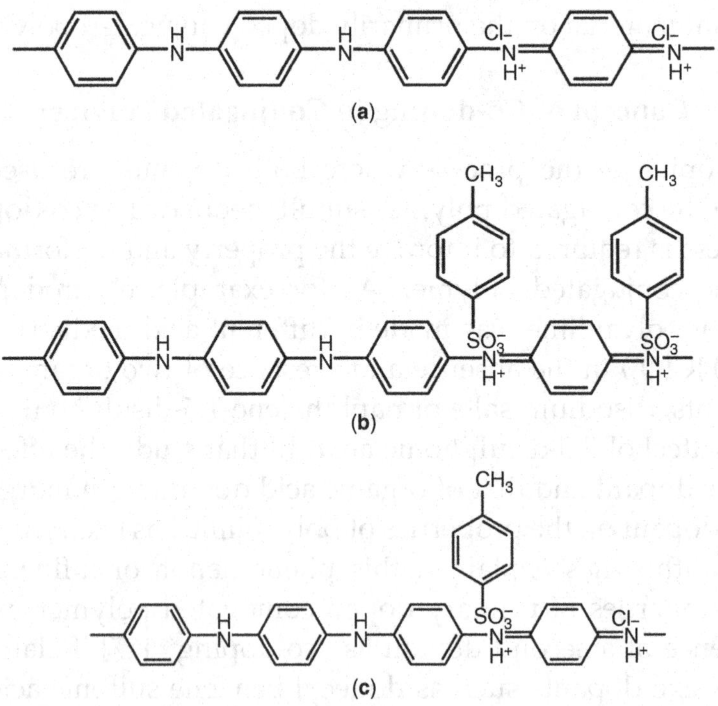

Scheme 1.3 Polyaniline (a) doped with HCl (b) doped with p-toluene sulfonic acid and (c) co-doped with HCl and p-toluene sulfonic acid [18].

simultaneous primary dopants that also are non-interacting with each other.

1.6 Doping as Probable Solution

The achievement of conjugated polymers for prospective application in various fields has not been widely recognized, especially by trade and business. As we have noted, these materials have not been highly successful in the marketplace compared to inorganic materials. This may be because they still have some disadvantages. The major problems for the use of conducting polymer are as follows:

a. Problems in synthesis
b. Problems in reproducibility
c. Problems in doping
d. Problems in processability
e. Problems in stability

Among these problems, a majority of conjugated polymers are suffering from doping problems, which are also associated with almost all the other disadvantages. We are already aware that the performance of doped conjugated polymers is greatly influenced by the nature of dopants and their level of distribution within the polymer. Yet another issue for doped conductive polymers at the present time is their relative instability under normal atmospheric conditions, which implies the need for better conductive polymers or better encapsulation, or both, when in a doped state. The electrochemical, mechanical, and optical properties of the doped conjugated polymers can be tailored by controlling the size and mobility of the dopants counter ions. The essential idea about the unusual nature of the species bearing charges, i.e., excited doped states of the conjugated systems, has been intensively discussed over the last twenty years. In this context the understanding of the nature of interaction by dopant with the π-conjugated systems is of foremost importance from the real application point of view. This rapid growth of interest in conjugated polymer-dopant interaction has been stimulated because of its fundamental importance to a cross-disciplinary section of investigators, chemists, electrochemists, biochemists, experimental and theoretical physicists, and electronic and electrical engineers. *Doping in Conjugated Polymers* is the first book dedicated to the subject and offers an A to Z overview. It details doping interaction, dopant types, doping techniques, influence of dopant on applications, etc. It explains how the performance of doped conjugated polymers is greatly influenced by the nature of dopants and their level of distribution within the polymer, and shows how the properties of the doped conjugated polymers can be tailored by controlling the size and mobility of the dopants' counter ions.

2

Classification of Dopants for the Conjugated Polymer

2.1 Introduction

Dopant is a generic name given to a vast number of materials which can have a doping effect on the conjugated polymer. These materials exist in countless numbers because of the very large variety of dopants for conjugated polymers. For generalization and better understanding of dopant interaction with the conjugated polymer it should be categorize. It is also helpful to evaluate the property of conjugated polymers doped with the same type of dopants. The property evaluation of conjugated polymers doped with the same type of dopants is very much necessary in developing the material for real use. Unfortunately the classification of dopant has been given the least importance by the research community and is also very confusing. The dopant type with preferable doping technique for some important conjugated polymers is shown in Table 2.1. Doping in conducting polymer can be distinctly classified as follows:

1. According to electron transfer
   a. p-type dopant
   b. n-type dopant
2. According to chemical nature
   a. Inorganic dopant
   b. Organic dopant
   c. Polymeric dopant
3. According to doping mechanism
   a. Ionic dopants or redox dopant
   b. Non-redox or neutral dopant
   c. Self-dopant
   d. Induced dopant

2.2 Classification of Dopant According to Electron Transfer

As the term doping is employed from inorganic semiconductor, the classification of dopant also can be employed according to the system. In inorganic semiconductor the doping occurs generally due to the incorporation of holes in the valence band, a p-type doping, or addition of electron in the conduction band, an n-type doping. Here, the doping occurs due to addition of p-type or n-type dopant in ppm level. Instead of the addition of electron or hole-like doping in inorganic semiconductor, the doping in organic conjugated polymers can be done through the charge transfer rection with the dopant resulting in partial oxidation or reduction of the polymers. The oxidizing or reducing agents are able to convert conjugated polymer to its salt form through the charge transfer reaction. It has been well accepted that the addition of a donor or an acceptor species to the conjugated polymer increases the conductivity of the polymer by several fold. The oxidation or reduction of various conjugated polymers for doping can be done by electron acceptor species like iodine, AsF_5, H_2SO_4, etc.,

Table 2.1 List of common dopant used for conjugated polymer with conductivity, doping technique and doping type

Polymer	Dopant	Counter ion	Conductivity (S/cm)	Used doping technique	Dopant type	Ref.
Polyaniline	HCl	Cl^-	12	Chemical (solution)	p-type, inorganic, neutral	33
Polyaniline	Tartaric acid	$C_4H_5O_6^-$	11	Chemical (solution)	p-type, organic, neutral	33
Polyaniline	poly(2-methoxyaniline-5-sulfonate) (PMSH)	PMS^-	13	Chemical (solution)	p-type, polymeric, neutral	33
Polyaniline	Dodecyl benzene sulfonic acid (DBSH)	DBS^-	300	Chemical (solution)	p-type, organic, neutral	33
Polyaniline	Acrylic acid	$C_3H_3O_2^-$	6.20×10^{-3}	In-situ	p-type, organic, neutral	26
Polyaniline	HCl	Cl^-	2.09×10^{-3}	In-situ	p-type, inorganic, neutral	26
Polyaniline	HCl	Cl–	200	In-situ	p-type, inorganic, redox	24
Polyaniline	Acetic acid	CH_3COO^-	4.21×10^{-2}	In-situ	p-type, organic, neutral	36
Polyaniline	Citric acid	$C_6H_7O_7^-$	80.43×10^{-2}	In-situ	p-type, organic, neutral	36

(Continued)

Table 2.1 (*Cont.*)

Polymer	Dopant	Counter ion	Conductivity (S/cm)	Used doping technique	Dopant type	Ref.
Polyaniline	Tartaric acid	$C_4H_5O_6^-$	24.53×10^{-2}	In-situ	p-type, organic, neutral	36
Polyaniline	Oxalic acid	$C_2HO_4^-$	18.17	In-situ	p-type, organic, neutral	36
Polyaniline	HCl	Cl^-	43.29×10^{-2}	In-situ	p-type, inorganic, neutral	22
Polyaniline	H_2SO_4	HSO_4^-	134.5×10^{-2}	In-situ	p-type, inorganic, neutral	22
Polyaniline	HNO_3	NO_3^-	94.59×10^{-2}	In-situ	p-type, inorganic, neutral	22
Polyaniline	$HClO_4$	ClO_4^-	25.37	In-situ	p-type, inorganic, neutral	22
Polyaniline	H_3PO_4	$H_2PO_4^-$	2.14×10^{-2}	In-situ	p-type, inorganic, neutral	22
Polyacetylene	I_2, Br_2, AsF_5	I_3^-, Br^-, AsF_6^-	10000	Chemical (gaseous)	p-type, inorganic, redox	24
Polyacetylene	HBr	Br^-	7×10^{-4}	Chemical (solution)	p-type, inorganic, neutral	46
Polyacetylene	HF	F^-	1.10	Chemical (solution)	p-type, inorganic, neutral	46

Polymer	Dopant	Counter ion	Conductivity (S/cm)	Used doping technique	Dopant type	Ref.
Polyacetylene	HCl	Cl^-	10^{-3}	Chemical (solution)	p-type, inorganic, neutral	46
Polyacetylene	Na, Li	Na^+, Li^+	10000	Electrochemical	n-type, inorganic, redox	24
Polyacetylene	HF	F^-	40	Chemical (solution)	p-type, inorganic, neutral	46
Polyacetylene	I_2	I_3^-	550	Chemical (gaseous)	p-type, inorganic, redox	23
Polyacetylene	AsF_5	AsF_6^-	1200	Chemical (gaseous)	p-type, inorganic, redox	23
Polyacetylene	$HClO_4$	ClO_4^-	50	Chemical (solution or gaseous)	p-type, inorganic, redox	23
Polyacetylene	R_4NClO_4 or $MClO_4$	ClO_4^-	50	Electrochemical	p-type, inorganic, redox	23
Polyacetylene	Li naphthanilide	Li^+	200	Chemical (solution)	n-type, inorganic, redox	23
Polyacetylene	Na naphthanilide	Na^+	25	Chemical (solution or gaseous)	n-type, inorganic, redox	23

(Continued)

Table 2.1 (*Cont.*)

Polymer	Dopant	Counter ion	Conductivity (S/cm)	Used doping technique	Dopant type	Ref.
Polyacetylene	Li	Li^+	10–100	Electrochemical	n-type, inorganic, redox	23
Polypyrrole	$R_4N\ BF_4$ or MBF_4	BF_4^-, ClO_4^-	500–7500	Electrochemical	p-type, inorganic, redox	24
Polypyrrole	p-toluenesul-fonic acid	tosylate	500–7500	Electrochemical	p-type, organic, redox	24
Polypyrrole	p-Methylben-zene sulfonic acid (MSAH)	MSA^-	16	In-situ	p-type, organic, neutral	27
Polypyrrole	p-Hydroxyben-zene sulfonic acid (HSAH)	HSA^-	11	In-situ	p-type, organic, neutral	27
Polypyrrole	p-Dodecylben-zene sulfonic acid (DBSH)	DBS^-	2	In-situ	p-type, organic, neutral	27
Polypyrrole	β-Naphthalene sulfonic acid (NSH)	NS^-	18	In-situ	p-type, organic, neutral	27

Polymer	Dopant	Counter ion	Conductivity (S/cm)	Used doping technique	Dopant type	Ref.
Polypyrrole	5-n-Butyl-naphthalene sulfonic acid (BNSH)	BNS^-	0.5	In-situ	p-type, organic, neutral	27
Polypyrrole	5-Sulfo-isophthalic sulfonic acid (SISH)	SIS^-	3	In-situ	p-type, organic, neutral	27
Polypyrrole	8-Hydroxy-7-iodo-5-quinoline sulfonic acid (HIQSH)	$HIQS^-$	3	In-situ	p-type, organic, neutral	27
Polypyrrole	Alizarin red acid (ARH)	AR^-	8	In-situ	p-type, organic, neutral	27
Polypyrrole	Camphor sulfonic acid (CSH)	CS^-	18	In-situ	p-type, organic, neutral	27
Polypyrrole	$CF_3SO_3R_4N$ or CF_3SO_3M	$CF_3SO_3^-$	150	Electrochemical	p-type, organic, redox	25,29

(Continued)

Table 2.1 (*Cont.*)

Polymer	Dopant	Counter ion	Conductivity (S/cm)	Used doping technique	Dopant type	Ref.
Polypyrrole	R_4NClO_4 or $MClO_4$	ClO_4^-	10	Electrochemical	p-type, inorganic, redox	25,29
Polythiophene	R_4NBF_4 or MBF_4, R_4NClO_4 or $MClO_4$, $FeCl_3$, $6H_2O$	BF_4^-, ClO_4^-, $FeCl_4^-$	1000	Electrochemical	p-type, inorganic, redox	24
Polythiophene	p-toluenesulfonic acid	tosylate	1000	Electrochemical	p-type, organic, redox	24
Polythiophene	$CF_3SO_3R_4N$ or CF_3SO_3M	$CF_3SO_3^-$	10–20	Electrochemical	p-type, organic, redox	25,29
Poly (3-methylthiophene)	$CF_3SO_3R_4N$ or CF_3SO_3M	$CF_3SO_3^-$	30–100	Electrochemical	p-type, organic, redox	25,29
Poly (3-methylthiophene)	R_4NPF_6 or MPF_6	PF_6^-	1	Electrochemical	p-type, inorganic, redox	25,29
Poly (3,4-dimethylthiophene)	$CF_3SO_3R_4N$ or CF_3SO_3M	$CF_3SO_3^-$	10–50	Electrochemical	p-type, organic, redox	25,29

Polymer	Dopant	Counter ion	Conductivity (S/cm)	Used doping technique	Dopant type	Ref.
Poly (3-alkylthiophene)	R_4NBF_4 or MBF_4, R_4NClO_4 or $MClO_4$, $FeCl_3$, $6H_2O$	BF_4^-, ClO_4^-, $FeCl_4^-$	1000–10000	Electrochemical	p-type, inorganic, redox	24
Polyphenylene-sulfide	AsF_5	AsF_6^-	500	Chemical (gaseous)	p-type, inorganic, redox	24
Polyphenylene-sulfide	CF_3COOH	$CH_3SO_3^-$	$<10^{-5}$	Chemical (solution)	p-type, organic, neutral	46
Polyphenylene-vinylene	AsF_5	AsF_6^-	1000	Chemical (gaseous)	p-type, inorganic, redox	24
Polyphenylene-vinylene	CH_3SO_3H	$CH_3SO_3^-$	10.7	Chemical (solution) or Electro-chemical	p-type, organic, neutral	46
Polyphenylene-vinylene	CF_3COOH	CF_3COO^-	$<10^{-7}$	Chemical (solution) or Electro-chemical	p-type, organic, neutral	46

(Continued)

Table 2.1 (*Cont.*)

Polymer	Dopant	Counter ion	Conductivity (S/cm)	Used doping technique	Dopant type	Ref.
Polyphenylene-vinylene	$FeCl_3$, $6H_2O$	$FeCl_4^-$	$<10^{-7}$	Chemical (solution) or Electro-chemical	p-type, inorganic, redox	46
Polythionylene-vinylene	AsF_5	AsF_6^-	2700	Chemical (gaseous)	p-type, inorganic, redox	24
Polythionylene-vinylene	CH_3SO_3H	$CH_3SO_3^-$	2.4	Chemical (solution) or Electro-chemical	p-type, organic, neutral	46
Polythionylene-vinylene	CF_3COOH	CF_3COO^-	$<10^{-4}$	Chemical (solution) or Electro-chemical	p-type, organic, neutral	46
Polythionylene-vinylene	$FeCl_3$, $6H_2O$	$FeCl_4^-$	5.3	Chemical (solution) or Electro-chemical	p-type, inorganic, redox	46
Polyphenylene	AsF_5	AsF_6^-	1000	Chemical (gaseous)	p-type, inorganic, redox	24

Polymer	Dopant	Counter ion	Conductivity (S/cm)	Used doping technique	Dopant type	Ref.
Polyphenylene	Li, K	Li^+, K^+	1000	Chemical (solution)	n-type, inorganic, redox	24
Polyisothia-napthalene	R_4NBF_4 or MBF_4, R_4NClO_4 or $MClO_4$,	BF_4^-, ClO_4^-,	50	Electrochemical	p-type, inorganic, redox	24
Polyfuran	$CF_3SO_3R_4N$ or CF_3SO_3M	$CF_3SO_3^-$	20–50	Electrochemical	p-type, organic, redox	25,29
Polyfuran	R_4NBF_4 or MBF_4, R_4NClO_4 or $MClO_4$,	BF_4^-, ClO_4^-,	100	Electrochemical	p-type, inorganic, redox	24
Polyazulene	R_4NBF_4 or MBF_4, R_4NClO_4 or $MClO_4$,	BF_4^-, ClO_4^-,	1	Electrochemical	p-type, inorganic, redox	24
Polyazulene	R_4NClO_4 or $MClO_4$	ClO_4^-	10^{-2}–10^{-1}	Electrochemical	p-type, inorganic, redox	25,29

or an electron donor species like alkali metals, etc. For example, an extensive list of the oxidizing and reducing agents for the doping of polyacetylene is available in the literature [19, 20]. By corellating with the concept of inorganic semicondutor doping the dopant for conjugated polymer can be similarly classified as p-type and n-type dopant.

2.2.1 p-Type Dopant

The p-type doping can be defined as the removal of electrons from the valence band, i.e., from HOMO of conjugated polymer by some oxidizing agent. The oxidizing agent acts as an electron acceptor or p-type dopant. So, p-type dopant oxidizes the conjugated polymer and leaves the polymer with a positively charged one (Scheme 2.1). As conjugated polymers are already electron rich due to having a π-system, a wide variety of common oxidizing agents have been shown to be effective for the doping of conjugated polymers. The common p-type dopants or oxidative dopants, which are usually electron-attracting substances, are Br_2, I_2, AsF_5, H_2SO_4, $HClO_4$, BCl_3, PF_3, SF_6, CH_3F, NO_2F, NO_2, $NO^+SbCl_6^-$, SO_3, $FeCl_3$, etc. Upon acceptor doping (p-type doping), an ionic complex is formed having positively charged conjugated polymer chains (e.g., polyacetylene) and counter anions (e.g., I_3^-, AsF_6^-, etc.). The p-doping by chemical method was first explained for trans-polyacetylene by oxidizing with iodine [20]. The trans-polyacetylene film was doped with ClO_4^- by this electrochemical p-type doping on cathode using propylene/$LiClO_4$ dopant electrolyte [19]. The p-type doping is possible for a large number of conjugated polymers like polypyrrole, poly(p-phenylene), polythiophene, poly(p-phenylene vinylene), poly(2,5-thienylene vinylene), polyaniline, polyaniline, polythiophene, etc.

2.2.2 n-Type Dopant

The n-type doping is the donation of an electron by a reducing agent to the empty conduction band, i.e., to the LUMO

Scheme 2.1 The p-type and n-type doping process in conjugated polymer.

of conjugated polymer. The reducing agent can be termed as donor or n-type dopant for the conjugated polymer. As for example, the n-type dopants are electron-donating substances like sodium naphthanilide, Na/K alloy, molten potassium, LiI, etc. In n-type doping the conjugated polymer is reduced and changes to a negatively charged one (Scheme 2.1). Due to donor doping (n-type doping), an ionic complex is formed having negatively charged conjugated polymer chains and counter cations (Na^+, K^+, etc.). Although p-type doping for conjugated polymer is easily possible by very common oxidizing agents, strong reducing agents like alkali metal are required for the doping of conjugated polymer, as conjugated polymers are already electron rich due to having a π-conjugated system. So, there are a relatively fewer number of reports available on the n-type doped conjugated polymers like alkali-metal-doped poly(p-phenylene) and poly(p-phenylene vinylene), polythiophene derivatives, and poly(3,4-ethylenedioxythiophene). The n-type doping was also demonstrated trans-polyacetylene by reducing with sodium amalgam or, preferably, sodium naphthanilide, or by electrochemical cathodic reduction of trans-polyacetylene film attached to a positive terminal electrode in $LiClO_4$ dissolved in tetrahydrofuran [19].

2.3 Classification of Dopant According to Chemical Nature

The fundamental chemical nature of the materials in chemistry is organic or inorganic. Generally, organic materials are made from (or extracted from) plants or animals and inorganic materials are made from rocks and minerals. The organic materials are mainly the compounds of carbon along with other elements which are found in living things and elsewhere. On the other hand, inorganic materials are materials which are composed of elements, minerals, or made from minerals, and are not animal or vegetable in origin. In another class, the polymer is a special type of organic or inorganic material in respect to its properties as well as versatility. In regard to size and molecular weight, in general, it is more than a million times bigger due to the repetition of part of a small molecule, i.e., mer or repeating unit. As we know, the properties of polymers generally depend on their chemical composition, molecular structure, molecular weight, molecular weight distribution and morphology. So, whatever be the doping interaction with the conjugated polymer, according to the above chemical nature of the reagents used for the doping of conjugated polymer, the dopant can be classified as inorganic, organic and polymeric dopant. In this doping class the properties of doped conjugated polymer vary basically due to the size of counter ions, which are in the order of inorganic dopant < organic dopant < polymeric dopant (Figure 2.1).

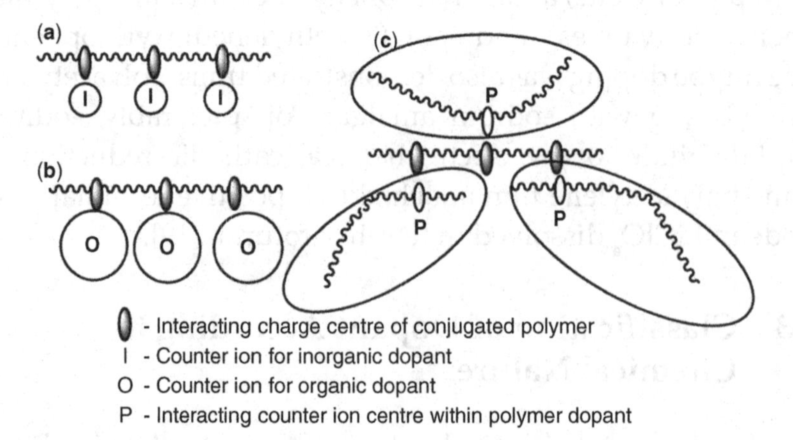

Figure 2.1 Arbitrary relative sizes of counter ions for (a) inorganic (b) organic and (c) polymeric dopant interacting with conjugated polymer chain.

2.3.1 Inorganic Dopant

As the name suggests, inorganic dopants are inorganic materials that are used to dope the conjugated polymer. The major compounds used as dopants for conjugated polymer are inorganic ones and they may be gaseous molecules (Halogens, oxygen, HCl, NO_x, AsF_3, etc.), metals (Li, Na, K, Mg, etc.), Lewis acids ($FeCl_3$, LiCl, $MgCl_2$, $SbCl_5$), inorganic acids (HCl, H_2SO_4, H_3PO_4, $HClO_4$, etc.), etc. Gaseous inorganic substances can be used for the doping of conjugated polymer directly in the gas phase or by dissolving them in suitable solvent. As for example, the oxidative doping of polyaniline is run either with chlorine or bromine in the gaseous phase or with gases like carbon tetrachloride in the organic solvent phase. One important added advantage for the inorganic dopant is that due to the small size of the dopant counter ion, the diffusion of the counter ions with the charged centers inside the polymer matrix is very high (Figure 2.1). For this reason generally inorganic-doped conjugated polymer shows better conductivity over the organic- or polymeric-doped conjugated polymer. The above statement is also true among the various types of inorganic dopants as well. That means the inorganic dopants with smaller counter ion are better in terms of conductivity of the doped conjugated polymer. As for example, polyaniline doped with HCl showed better conductivity than polyaniline doped with H_2SO_4, and H_3PO_4 due to smaller counter ion Cl^- [21]. However, the conductivity not only depends on the size of counter ion for inorganic acid dopant but also on the oxidizing nature of the acid used. Polyaniline (emeraldine salt) was synthesized in the presence of different inorganic protonic acid dopants, e.g., hydrochloric acid (HCl), nitric acid (HNO_3), perchloric acid ($HClO_4$), sulfuric acid (H_2SO_4), and hydroiodic acid (HI). The $HClO_4$-doped polyaniline showed the highest conductivity (109.04 S/cm) due to more oxidizing property, while HI-doped polyaniline gave the lowest conductivity of 0.02 S/cm [22]. The smaller counter ion of inorganic dopant can also favor the better form of doped conjugated polymer to show better conductivity for the polymer. This phenomenon can be well explained for iodine-doped cis-polyacetylene [19], where the cis-form is converted to trans-configuration. Other inorganic dopants investigated for doping of polyacetylene are SbF_6^-, AsF_6^-, $C1O_4^-$, $FeC1_3$, $IrC1_6^{2-}$, etc. The above dopants are also found suitable to dope the other conjugated polymers like polythiophene, polypyrrole, polyphenylene, and polyphenylene vinylene (Table 2.1) and their derivatives [23–28]. The inorganic dopant, which is basically a good electrolyte due to its easy ionizable nature, can be used to dope the conjugated polymer by the electrochemical method. Here, the dopant itself can also serve as supporting electrolyte for the process. As for example, the doping effect of supporting inorganic electrolytic salt MX ($M^+ = Li^+$, $N(Bu)_4^+$; $X^- = C1O_4^-$, BF_4^-, PF_6^-, $CF_3SO_3^-$) for various polythiophene derivatives electrochemically deposited on Pt has been investigated [29]. An attempt has been made to electrochemically prepare polyaniline, poly(o-toluidine) and their copolymer thin films doped by several inorganic salts viz., K_2SO_4, Na_2SO_4, Li_2SO_4, $MgSO_4$, KCl, NaCl, LiCl and $MgCl_2$ using aqueous solution of the salt with H_2SO_4 as electrolyte. In the overall study, an increase in conductivity is observed for all above-mentioned dopants, and among these K_2SO_4 is found to be the best in the sulfate category and KCl in the chloride category [30, 31]. The inorganic dopant having a weak electrolytic nature is able to dope the conjugated polymer chemically in organic solvent medium as well. The weak electrolytic molecules, e.g., $(NO)^+(PF_6)^-$, $FeCl_3$, $SnCl_4$, etc., are used for oxidation doping of polyaniline by solution doping technique in suitable organic solvents like chloroform, carbon tetrachloride, etc. [32].

The most important problem for conjugated polymer doped with inorganic dopant is its poor environmental stability due to the hydrophilic nature of the dopant. The conductivity of the inorganic-doped polymer decreases sharply with time due to interaction with moisture present in the environment. Based on gravimetric measurements, the effect of water on thin films of polyaniline doped with HCl, H_2SO_4, and H_3PO_4 has been studied. HCl dopant showed a rapid release of dopant ions with time, however, H_3PO_4 and H_2SO_4 dopant showed a slow release of dopant ion upon water treatment of the doped polyaniline film [21]. Conjugated polymers doped with inorganic dopants are generally insoluble in organic solvents and thus solution processability of the polymers in doped form is not processable. Therefore, it requires treatment with a base to remove the dopant for making film from suitable solution, and then it is again doped with the same or different dopant to obtain better conductivity characteristics. For this problem, and also for the poor film-forming nature of the rigid conjugated polymer, doped polymer pellet is generally used to get the conductivity, preferably by the four-probe or two-probe method. The electrochemically deposited film of conjugated polymer on the electrode surface is one of the solutions for the problem. Another solution is after processing the conjugated polymer doped with organic dopant the dedoping/redoping can be achieved by inorganic dopant. The dedoping/redoping study after processing polyaniline doped with dinonyl naphthalene sulfonic acid has been reported for various inorganic dopants like HCl, HNO_3, H_3PO_4 and H_2SO_4 [33].

2.3.2 Organic Dopant

The delocalized π-electronic structure of conjugated polymer leads to large electronic polarizability and a rigid polymer backbone, which is insoluble in most of the solvents. High electronic polarizability leads to a large interchain π-π attraction (dispersive force), which favors aggregation rather than solvation. Doped conjugated polymers are composed of hydrophilic inorganic dopants complexed with the polymer chain, which also contains hydrophobic organic segments. It is difficult to find solvents to solvate simultaneously both hydrophilic and hydrophobic segments in one chain. Organic dopants are organic chemicals which are used to dope the conjugated polymer to improve the solubility of the doped conjugated polymer in common organic solvent. The organic dopant having hydrophobicity or much less hydrophilicity also improves the environmental stability of the doped complex to some extent. For organic dopant the counter ion is big inside, so the diffusion of the counter ions with the mobility of the doped centers inside the polymer matrix is restricted or localized (Figure 2.1). The polypyrrole films doped with various sulfonate derivatives of anthraquinone, naphthalene, and copper phthalocyanine are examples for the above statement [34]. Here, the conductivity of the doped polypyrrole decreases with an increase in the number of sulfonate groups due to localization of positive charge carriers within polypyrrole films. To improve the solubility of conjugated polymer in the doped form various organic dopants are used with big counter ion. The big counter ion of organic dopant restricts the close packing of doped conjugated polymer chain. Thus the use of this big dopant increases the processability of the doped polymer. Polyaniline doped with organic protonic acid viz., acetic, citric, oxalic, and tartaric acid has been reported soluble in various common organic solvents like m-cresol, carbon tetrachloride, dimethyl sulfoxide, etc. The organic substances are very weak electrolyte but electrochemical doping can also be employed to dope the conjugated polymer. Quarternary ammonium salts of the type R_4NX (where R = Alkyl, Aryl group and X = Cl^-, Br^-, I^-, ClO_4^-, BF_4^-, PF_6^-, $CF_3SO_3^-$, $CH_3C_6H_4SO_3^-$) are commonly used as supporting organic electrolytes. Electropolymerized films of polyaniline were doped by electrochemically driven anion exchange with indigo tetra-sulfonate, hydroquinone-2, 5-disulfonate and 1,2-naphthoquinone-4-sulfonate by simple potential cycling, provided in an acid solution containing the corresponding dopants [35].

Due to having bigger counter ion, which restricts the charge mobility inside the polymer, the same conjugated polymer doped with organic dopant generally shows less conductivity than the polymer doped with inorganic dopant. Despite having higher conductivity for conjugated polymer doped with inorganic dopant, the use of organic dopant to dope conjugated polymer is of great interest for improvement of the solution processability of the doped conjugated polymer with organic dopant. Thus the use of this big dopant increases the processability of the doped polymer but it decreases the conductivity of the polymer as well. Polyaniline doped with organic protonic acid viz., acetic, citric, oxalic, and tartaric acid has been reported. In accordance with these results the conductivity is also found to be higher in oxalic acid doped material due to higher acid strength and moderate size of counter ion [36]. Here, the organic acid dopant with higher acid strength and moderate size of counter ion is better dopant for the polyaniline in terms of better conductivity and moderate solubility in organic solvent. Another problem for organic dopant is that the film casting is very difficult from the solution of doped conjugated polymer. This is often due to the low molecular weight resulting from the organic acid doped conjugated polymer during synthesis. In one approach, the organic dopant with suitable bigger counter ion can be used first for film casting and then dedoping/redoping can be done to achieve better conductivity. As for example, polyaniline doped with dinonyl naphthalene sulfonic acid is highly soluble in organic solvents. After making the film the polymer can be dedoped by NaOH and again redoped by simple solution doping technique by using various organic dopants like p-toluene sulfonic acid, tartaric acid, oxalic acid, dodecyl benzene sulfonic acid, and 3-methyl aniline sulfonic acid [33]. The dedoping of the above system also occurred upon extended immersion in neutral methanol due to simple dissolution of the counter ion, as solubility of organic counter ion is very high in organic solvent.

2.3.3 Polymeric Dopant

Polymeric dopants are easily ionizable functionalized polymer electrolytes which can be used as dopants for conjugated polymers. The purpose of using a polymeric dopant is to avoid the easy removal of a small molecule dopant to increase the stability of the doped complex. In order to overcome the problem of conductivity loss in moisture or heat due to small inorganic or organic counter ion, one of the research trends is towards the formation of polymeric dopant composite with the conjugated polymer. For example, polyaniline-polyacrylic acid composite thin films were deposited at room temperature on glass and polymethyl methacrylate substrates [37]. Generally, doping is an acid-base reaction and thus polymer with some acid property can act as a dopant for conjugated polymer, which is basic in nature like polyaniline. The doping behaviors of polymeric acid dopant, e.g., polystyrene sulfonic acid and poly(2-acrylamido-2-methyl-propane sulfonic acid) on polyaniline were demonstrated [38]. The FTIR spectral characterization confirmed the presence of the acid-base reaction product between polyaniline and polymeric dopant incorporation into the matrix. Polymeric dopant has an added advantage over inorganic or organic dopant in that it often improves the film-forming properties with good mechanical strength for the conjugated polymer, provided the dopant polymer itself should have good film-forming property. Since the films of doped conjugated polymer are necessary from the practical application point of view for better random molecular structure, polymeric dopants have drawn attention. The dedoping/redoping was possible with various polymeric dopants like polyvinyl phosphate and poly(2-methoxyaniline-5-sulfonate) after processing polyaniline doped with dinonyl naphthalene sulfonic acid [33]. Due to similar physical properties of the polymeric dopant and conjugated polymer, the association of both the compatible materials is possible in solid or solution or gel state. Polystyrene with different degrees of sulfonation was employed as a polymeric dopant for polyaniline by three different methods: in solid state, in solution and in gel state [39]. The flexibility is also there for polymeric dopant, which is introduced in the conjugated polymer matrix by simple blending of both the polymers. As for example, polyaniline doped with poly(ethane sulfonic acid), poly(styrene sulfonic acid), polyacrylic acid, poly(2-(acryl amido)-2-methyl-1-propane sulfonic acid), poly(amic acid), etc., can be prepared by the simple solution blending technique [40–42]. Polymeric doping can be chemically done by simultaneous synthesis of both polymers using the same reagent with or without the presence of other dopant. Here, the other dopant can be used as co-dopant to enhance the property of the conjugated polymer. Electrically-conducting polypyrrole/poly(2-acrylamido-2-methyl-1-propane sulfonic acid) dopant composites were chemically synthesized in an aqueous solution containing dodecyl benzene sulfonic acid. The electrical conductivity of the polypyrrole/poly(2-acrylamido-2-methyl-1-propane sulfonic acid)/dodecyl benzene sulfonic acid film having excellent mechanical property is close to that of polypyrrole doped with only dodecyl benzene sulfonic acid. Here, poly(2-acrylamido-2-methyl-1-propane sulfonic acid) is acting as a co-dopant for the composite [43].

The main problem for polymeric dopant is interaction with the conjugated polymer is very weak. This is because the reactivity of the ionizable functional group within the polymeric dopant is much less. After doping, here the mobility of contour ion is also a big problem as diffusion of big polymer dopant counter anion within the conjugated polymer is restricted (Figure 2.1). Therefore, conjugated polymer doped with polymeric dopant generally has very low conductivity compared with that of an inorganic- or organic-doped one. Here the better reactivity or ionizability of the polymer dopant is also required for better doping in the conjugated polymer. In the case of functionalized polymer, it is very difficult to behave as very good electrolytes in solution. So, there is almost no report for this type of doping by electrochemical method. Such a rare example is when poly(styrene sulfonate-co-vinyl ferrocene) is used to dope polypyrrole during electrochemical polymerization in the presence of the dopant [44]. The mixing or blending of conjugated polymer with the polymeric dopant is a very big problem as it is only possible when both the polymers are soluble in a single solvent. Moreover, compatibility of conjugated polymer with the polymeric dopant is also a problem.

2.4 Classification of Dopant According to Doping Mechanism

This classification is quite confusing in the first category, as the electron transfer between the dopant and conjugated polymer can be comparable with the doping mechanism. From the previous discussion it is clear that the aim of doping is to generate the charge center stabilized by the charged counter ion within the conjugated polymer. Now, within the conjugated polymer the charge center can be generated by transfer of electron from the conjugated polymer or by adding some charged species to the conjugated polymer. Again, the process can be possible for a group present within the conjugated polymer itself as side chain. Even so, a complete transfer of electron is not necessary from doping, i.e., slight shifting of labile π-electron cloud should be sufficient for doping within a conjugated polymer. Therefore, the doping mechanism is different with varying pathways for doping having similar or different reaction of dopant with conjugated polymer, i.e., charge transfer. The types of dopant according to the doping mechanism will be discussed in following sections.

2.4.1 Ionic Dopant or Redox Dopant

Ionic dopants are oxidizing or reducing agents which are responsible for the electron transfer process with the conjugated polymer during doping. The counter ion of the dopant remains with the polymer (as a salt) to make it neutral. It is also well established that in redox doping the exposure of conducting polymer (e.g., polyacetylene) to an oxidizing agent (e.g., I_2, Br_2, AsF_5 etc.) leads to the reduction of conjugated polymer. On the other hand, the exposure to an reducing agent (e.g., Na, Li, etc.) leads to oxidation of the conjugated polymer. So, here the conjugated polymer salt is formed due to the coexistence of conjugated polymer ion with the corresponding dopant ion, and only the electron transfer occurs between them. The redox doping mechanism for an arbitrary conjugated polymeric system is shown in Scheme 2.2. Poly(acetylene), poly(p-phenylene), polyeheterocyclics like poly(thiophene), poly(pyrrole), poly(furan) and their derivatives with no strong basic centers in their backbone are preferable for undergoing redox doping. All conducting polymers and their derivatives, e.g., poly-(p-phenylene), poly(phenylene vinylene), polypyrrole, polythiophene, poly(furan), poly(heteroaromatic vinylenes), polyaniline, etc., undergo redox doping by chemical and/or electrochemical processes during which the number of electrons associated with the polymer backbone changes.

Doping involving no dopant ions in photo doping or charge injection doping can also be preferably termed as redox doping [45]. For example, photo-doping of trans-polyacetylene in radiation of energy greater than its band gap leads to simultaneous oxidation and reduction of part of the polymer chain. Under appropriate experimental conditions solitons can be observed with the electrons and holes separate within the polymer chain. Similarly charge-injection doping is also redox doping due to the formation of charged soliton during doping of the conjugated

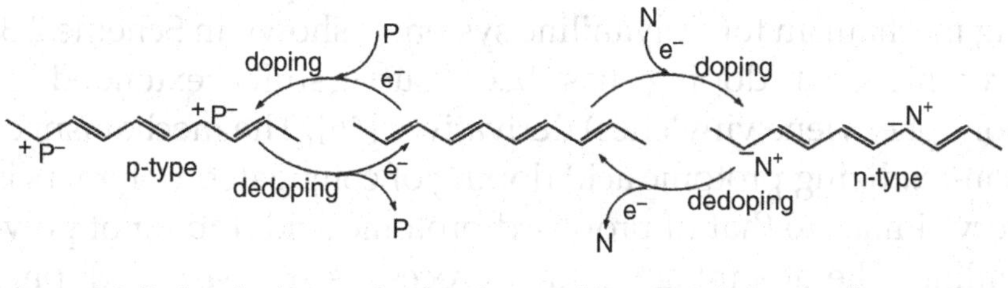

Scheme 2.2 Redox doping mechanism for an arbitrary conjugated polymer system.

polymer. In charge-injection doping application of an appropriate potential across metal/insulator/semiconductor (MIS) (here conjugated polymer) configuration can give rise to a conjugated polymer charge layer, i.e., doped conjugated polymer layer with higher conductivity. For example polyacetylene or poly(3-hexylthiophene) are used in the MIS device in doped form by this method without any associated dopant ion [45].

2.4.2 Non-redox Dopant or Neutral Dopant

Neutral dopants produce negative and positive ions simultaneously for doping conjugated polymers. In non-redox doping the energy levels are rearranged for the conjugated polymer during doping without changing the number of associated electrons. However, here also the oxidation or reduction (i.e., p-type or n-type) of conjugated polymer occurs due to the addition of positive or negative ion with the conjugated polymer and the complexion of this charged polymer with counter ion results in the doped polymer salt. Thus the doping is termed as neutral doping, as no electron transfer occurs from or to the conjugated polymer. Protonic acids as non-redox dopants have been found to be effective for many of the conjugated polymers. Non-redox doping is the preferable doping for strong basic centers containing conjugated polymer like polyaniline. For example, all the above cited literature of protonic inorganic, organic or polymeric acid doping to polyaniline and its derivatives is non-redox dopant. In that case the resulting polyaniline salt is due to a combination of protonated positively charged polyaniline and the counter anion of the corresponding acid. Generalized non-redox protonic acid doping mechanism for polyaniline system is shown in Scheme 2.3. Protonic acid doping has been successfully extended to poly(phenylene vinylenes) derivatives [46]. The mechanism of non-oxidizing protonic acid doping of conjugated polymers is very similar to that of proposed protonic acid doping of polyaniline. The advantage of this process is the easy dedoping

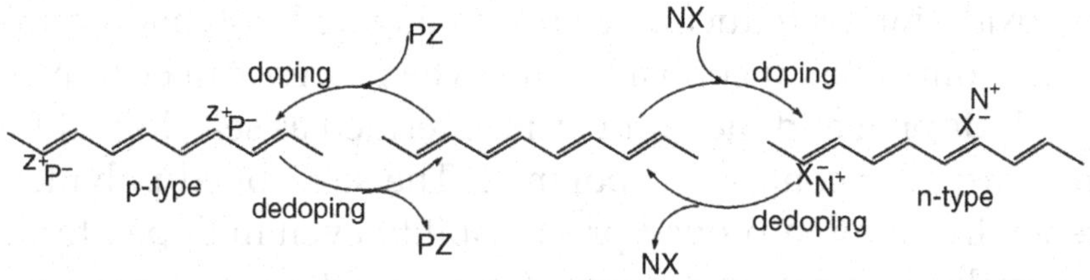

Scheme 2.3 Non-redox or neutral doping mechanism for an arbitrary conjugated polymer system.

and redoping process, which is particularly important for the processing purpose of the conjugated polymer. The chemically synthesized polyaniline salt requires treatment with a base to get the undoped polymer, and then after film casting from N-methyl-2-pyrrolidone (NMP) the polyaniline film can be redoped with the acid. Also, protonic acid doping of films of polyacetylene, poly(p-phenylene vinylene), poly(p-phenylene sulfide), and poly(3-dodecyl-thiophene) is possible. The doped polypyrrole with dodecyl sulfate as a counter-ion shows a very high conductivity 10^2 S/cm [47]. However, the non-redox or neutral doping is particularly important for polyaniline and its derivatives.

2.4.3 Self-dopant

As discussed above, in the case of redox or non-redox doping, generally the counter ions of dopants is required to preserve charge neutrality with the charge center of doped conjugated polymer. The restriction of mobility of large counter ions during the conduction limits many of the important characteristics like electrochromic switching, charging rates, stability, solubility, etc. Here, the self-doped conjugated polymer is introduced, in which there is an ionizable functional group covalently attached with the conjugated polymer chain acting as a dopant for the polymer. The ionizabe functional group can donate smaller ion-like proton to the neighboring center of the polymer, and then this charged functional group itself maintains charge neutrality for the center. Thus, in self-doping the oxidation or reduction of the conjugated polymer occurs by the ionizable functional groups. The process can be termed as self-doping and the group can be termed as self-dopant for the particular conjugated polymer. The self-doped polymers generally show improved processability even in doped form, especially in aqueous solvent due to zwitter-ionic structure. However, water solubility of the self-doped conjugated polymer often becomes a real problem for the purpose of practical application as moisture can cause serious damage in the device. The book written by Freund and Deore can be used as a reference on self-doped conducting polymer [48]. Sodium salts and acid forms of poly(3-thiophene ethane sulfonate) and poly(3-thiophene butane sulfonate) prepared by electropolymerization were the first reported self-doped conducting polymer (Scheme 2.4) [49, 50]. The polymers had good water solubility in neutral undoped (insulating) state as well as in doped (conducting) state. The strong evidence for self-doping is that the monomer itself can act as the electrolyte during electrochemical polymerization of potassium salt of propyl sulfonate thiophene derivative [51]. Chemical synthesis of self-doped conjugated polymer is also reported by oxidizing sodium 3-(3′-thienyl) propane sulfonate monomer in aqueous ferric chloride [52]. This approach of introducing flexible ionic substituents has been successfully extended to polypyrrole,

$SO_3^- M^+$ $(CH_2)_3$

M = H, Na, Li *etc.*
M = SO_3

Poly (3-thienyl) propanesulfonate derivative

Sodium poly (3-pyrrole butanesulfonate)

Sulfonated polyaniline derivative

Sodium salt of sulfonated polyniline

Scheme 2.4 Structure of some important self-doped conjugated polymers [49, 50].

polyaniline, polyphenylenes and polyphenylene vinylenes. Though the conductivity of self-doped ring-sulfonated polyaniline (0.1 S/cm) is significantly lower than that of sulfuric acid-doped polyaniline (1–10 S/cm), it is independent of pH up to 7 due to the availability of a high concentration of protons for doping in the vicinity of the polymer [53].

2.4.4 Induced Dopant

Induced doping is the term used where no oxidation or reduction occurs for conjugated polymer. In the case of induced doping only the shifting of electron cloud occurs due to the strong ionic nature of the functional groups or some external species. These functional groups or external species can be termed as induced dopant for the conjugated polymer. It cannot be differentiated as p-type or n-type dopant because there is no electron transfer (redox) and no addition of extra ion with counter ion (non-redox) is observed. Although it is like self-doping, it is better to explain it as induced doping by sodium cations attached with the free –OH groups of the poly(m-aminophenol). The difference from self-doping is that oxidation or reduction is not observed for the conjugated polymer by the ionizable functional groups. Due to the ionic nature of functional groups such as $-O^{-}Na^{+}$, which results in strong electron donation to the π-electron system (Scheme 2.5), the polymer shows a moderate conductivity 2.34×10^{-5} S/cm. This induced donation of electron can be confirmed from FTIR spectra, as the C–N or C–O bond with attached C=C bond may be able to get some C=C=N or C=C=O bond character [54]. Similarly, the silver(0) nanoparticle shows some doping effect on the same polymer in poly(m-aminophenol)/silver nanocomposite having the highest DC-conductivity of 1.03×10^{-6} S/cm. Here, the flow of electron delocalization in the polymer chain is increased as silver nanoparticles are trapped by the free –OH groups of the polymer. Although it is not mentioned, it can also be considered as induced doping by the

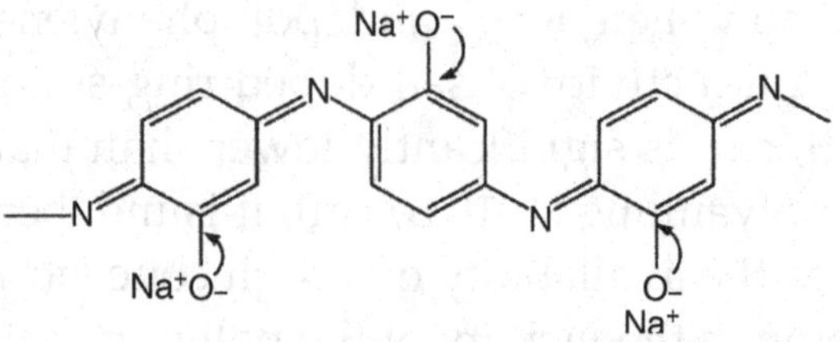

Scheme 2.5 Induced doping by sodium ion in poly(m-aminophenol). Reproduced with permission from ref. [54], Copyright © 2010 Elsevier B.V.

silver nanoparticle which attracts the π-electron cloud of the conjugated poly(m-aminophenol) [55]. Probably due to the same reason, the polyaniline/silver nanoparticles composite, synthesized by the reduction of silver nitrate using different acid-doped polyaniline salt, shows good conductivity. From FTIR analysis it has been confirmed that only the residual emeraldine units remain protonated by nitric acid, which is a reaction by-product. Therefore, the very high conductivity of the polyaniline/silver nanocomposite containing 24–27% silver nanoparticle with few acid-doped units can be explained by "induced doping," although this is not considered by the author [56]. The induced doping effect of micro- and nano-copper particles on polypyrrole has been investigated. The authors have observed the doping effect of copper particles in their zero state on polypyrrole that can be also termed as induced doping [57].

3

Doping Techniques for the Conjugated Polymer

3.1 Introduction

The electrical properties of conjugated polymers, and hence their performance as semiconducting materials, strongly depend on doping. The accurate doping of conjugated polymer to achieve the highest conductivity is a great challenge for the material scientist. Therefore, the doping process is considered one of the most important issue in this field. Before discussing doping techniques, it is important to know in which form the conjugated polymer can be used for doping. Consideration of the matter falls out of the scope for discussion, only the names of the processes can be prescribed according to the literature, e.g., dip-coating, solvent evaporation or solution casting, spin coating, layer-by-layer (LBL) self-assembly technique, Langmuir-Blodgett, electrochemical polymerization, deposition by radio frequency polymerization, polymer deposition followed by crosslinking, pellet preparation, vacuum deposition, etc. The preferable doping technique for the doping of some important conjugated polymers has already been shown in Table 2.1. Now, doping in the conjugated polymer may be carried out mainly by the following five techniques:

1. Electrochemical doping
2. Chemical doping
3. *In-situ* doping
4. Radiation-induced doping or photo doping
5. Charge-injection doping

3.2 Electrochemical Doping

For use in a wide variety of electrical and electronic device applications, it is highly desirable to dope the conjugated polymer by the electrochemical doping technique. The highly reversible doping/dedoping nature of the electrochemical process makes it more advantageous over the chemical process. As for example, the electrochemical doping/dedoping process is very helpful in understanding the performance of conjugated polymer as an important material in rechargeable batteries. Furthermore, for the use of doped conjugated polymer in semiconductor device applications the p-type and n-type electrochemical doping may have a correlation with the p-n junction properties. In electrochemical doping, the electrode supplies the redox charge (by donating or withdrawing electron) to the conjugated polymer, and ions diffuse from the nearby electrolyte into (or out of) the polymer to compensate the charge. Another reason for the popularity of the classical electrochemical technique in the conducting polymer community is that with electrochemical characterization it is possible to have important information. As for example, the amount of charge used up in the doping reaction, which is merely electrochemical oxidation or reduction, determines the degree of doping. The electrochemical set-up may be a standard three-electrode system consisting of a working electrode, counter electrode and reference electrode (Figure 3.1), or a two-electrode system

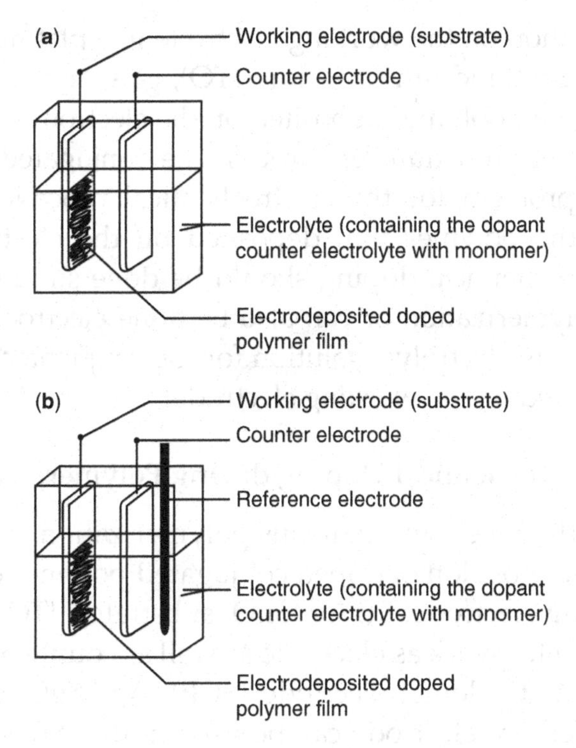

Figure 3.1 The (a) two electrode system and (b) three electrode system for electrochemical doping.

(without reference electrode). The three-electrode system is preferred over the two-electrode system as proper electrochemical characterization is possible in the three electrode set-up. Upon electrolysis, the electrolyte is dissociated and dopant counter ion migrates to the oxidized or reduced conjugated polymer, which is attached on the working electrode. Electrochemical p-type doping can be done by anodic oxidation by immersing a conjugated polymer on negative electrode within dopant electrolyte. The electrochemical n-type doping, generally by a metal ion, can be done using the metal as cathode material immersed within the metal cation containing electrolyte. Electrochemical doping allows precise control of the doping level by the applied voltage and leads to a homogeneous distribution of dopants in bulk. The electrochemical doping can be performed using suitable electrolyte during the electrochemical polymerization on the working electrode like platinum, stainless steel, gold, indium tin oxide (ITO), glass, etc. Otherwise, the conjugated polymer deposited on the electrodes is directly used. The electron transfer through the conjugated polymer creates a problem for the electrochemical process and also damages the polymer film deposited on the electrode surface. Electrochemical doping should be done simultaneously during polymerization or can also be done electrochemically in appropriate electrolyte solution for the conjugated polymer film deposited on the working electrode.

3.2.1 Electrochemical Doping during Polymerization

In this method, i.e., simultaneous polymerization and doping the electrodeposition of doped conjugated polymer occurs on the working electrode. Pt, Au, SnO_2 substrates, ITO, stainless substrates, etc., work as electrode as well as counter electrode, and saturated calomel electrode (SCE), Ag/AgCl electrode, etc., as reference electrode can be used in the three-electrode system. Generally ionic-type supporting electrolyte, which is used in the polymerization medium, helps in the electrochemical process by supplying ions as well as by doping the conjugated polymer by coupling ionically with monomer. As a wide choice of the supporting electrolytes is possible, the electrochemical polymerization method offers a large variety of dopant ions. This technique has been used to prepare stable, flexible films of polypyrrole [58, 59]. Poor nucleophilic aprotic solvents (e.g., acetonitrile, benzonitrile, etc.) are preferably used. Certain aprotic solvents with some nucleophilic character, such as dimethyl formamide, dimethyl sulfoxide, hexamethyl phosphoramide and hydroxylic, are also used by reducing the nucleophilicity with an addition of suitable protic acid. Quarternary ammonium salts of the type R_4NX (where R = Alkyl, Aryl group and X = Cl^-, Br^-, I^-, ClO_4^-, BF_4^-, PF_6^-, $CF_3SO_3^-$, $CH_3C_6H_4SO_3^-$) are commonly used supporting electrolytes that are soluble in aprotic solvent [60]. Electropolymerization of 3,4-Ethylenedithiothiophene was carried out in 0.005 M solutions of the monomer in a 0.1 M environment of four different electrolyte salts: $(Bu)_4NClO_4$, $(Bu)_4NBF_4$, $(Bu)_4NPF_6$ and $(Bu)_4NCF_3SO_3$ in CH_3CN. The counter ion of the electrolyte salt were used as a dopant for the synthesized polymer and the doping ability was correlated to the base strength of these anions $ClO_4^- > BF_4^- > PF_6^- > CF_3SO_3^-$ [61].

3.2.2 Electrochemical Doping after Polymerization

In the other method, the neutral polymer is first synthesized by conventional ways and subsequently treated electrochemically with a strong oxidant/reductant in order to produce the doped conjugated polymer. Thus, the method is flexible to the conjugated polymer synthesized by either of the methods, chemically or electrochemically. The conjugated polymer deposited electrochemically on the suitable electrode, as described earlier, can be used for further doping by this process. However, the chemically synthesized soluble conjugated polymer is attached on the suitable electrode by a solvent-based technique such as solution casting (preferable), dip coating, spin coating, etc., for the purpose of electrochemical doping. During the doping process within some electrolyte solution, the mechanical stability of the conjugated polymer-coated electrode is a big problem due to weak adhesion of conjugated polymer on the electrode. Polyacetylene can be oxidized or reduced electrochemically using a variety of electrolyte solutions. This method is not very popular, as electrochemical *in-situ* doping is easy, convenient and a single-step method. Moreover, chemical doping is an easy and straightforward method for electrochemically synthesized undoped conjugated polymer.

3.3 Chemical Doping

The chemical doping process is an efficient as well as straightforward process and it is also very useful for both chemically- or electrochemically-synthesized conjugated polymer. For this doping, the film or the powder of chemically-synthesized conjugated polymer can be used successfully. The chemical doping may be performed directly for electrochemically-synthesized conjugated polymer on the electrode surface, but the electrode material must be inert within the dopant environment. Otherwise, the polymers deposited on the electrode surface are peeled off as self-standing films for chemical doping. However, the main problem is that in most cases the polymer-coted electrode should be used directly, as the mechanical stability for ultrathin, peeled-off film from the electrode surface is very poor. The pellet preparation technique is very useful for polymers, especially those which are not soluble in organic solvent. In this technique the well-dried conjugated polymer powder is made into a thin pellet of particular thickness using a steel die in a hydraulic press under a certain pressure. Generally the pellet should be made after doping the polymer powder, as exposure of the particle for pellet is less and also shows poor mechanical stability in dopant solvent during the doping process. However, in the chemical doping technique control over the doping level is very poor and complete doping to the optimum concentrations often results in inhomogeneous doping. Chemical doping can be done by a gaseous or solution pathway as described below.

3.3.1 Gaseous Doping

In gaseous doping, generally polymers are exposed to the vapors of the doping agent inside a closed chamber. So, in this heterogamous phase reaction, solid conjugated polymer is reacting with the gaseous dopant. The gaseous doping set up is shown in Figure 3.2. To get the highest conductivity for the doped conjugated polymer at optimum doping state, the temperature and pressure of the chamber can be maintained. The temperature can be helpful in enhancing the interaction of dopant gas with the conjugated polymer or in producing the vapor, especially from sublimated solid iodine [62]. At reduced pressure the concentration of the vapor increases within the closed chamber and as a result the optimum doping of the conjugated polymer requires less time. Heating is generally avoided in this method, as the loosely adsorbed dopant counter ion is due to the escape from the surface of the film, which results in the dedoping of the conjugated polymer. Atmospheric pressure is generally preferred over the application of particular pressure, as the extra complicated set-up is required to control the optimum pressure. Moreover, the time required for optimum doping even at room temperature and pressure by the gaseous doping process are very low, within the range of 10 min to 3 h. The polyacetylene samples were doped to maximum conductivity by exposure to iodine vapor at pressures below 1 torr and by using a 5-mM solution of iodine in hexane. The time required to attain the maximum conductivity in vapor (pressure of 1 atm at room temperature) was only 3 h, while for solution-phase doping it was approximately 8 h [63]. The conventional gaseous doping method is useful to dope polyaniline, polyacetylene, polypyrrole, etc., by iodine, bromine or chlorine vapor. For example, Br_2 vapor or AsF_5 gas gives oxidized polyphenylene [64], and solid acetylene reacts with AsF_5 to give conducting polyacetylenes [65]. Poly(*p*-phenylene), poly(2,6-naphthylene), poly(2,7-naphthylene), poly(1,5-naphthylene), and poly(1,4-naphthylene) were doped with SbF_5 vapor by using this method [66].

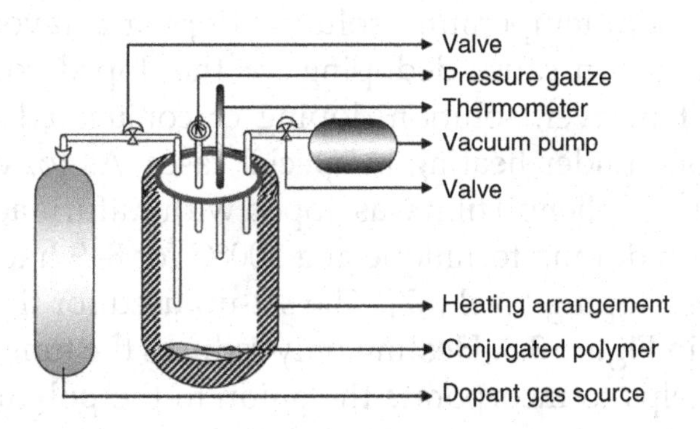

Figure 3.2 Gaseous doping set-up for conjugated polymer.

3.3.2 Solution Doping

In the popular solution doping method a solvent is used in which the dopant and all the products of dopant during doping are soluble but the polymer is not soluble. So, this method is also a type of heterogamous phase reaction. By this method a suitable part of the film or powder of the undoped conjugated polymer is dipped in a solution of the dopant in which the polymer is not soluble. Polyacetylene is doped with sodium ion by solution doping technique from sodium napthanilide solution. Solution phase n-Doping can be done for transpolyacetylene by treating with liquid sodium amalgam or, preferably, sodium in naphthaline solution [19, 20]. Polymers like polyaniline, polypyrrole, polythiophene, etc., are also used to dope with various inorganic acids by solution doping technique. H_2O_2 only oxidizes the polyaniline within the aqueous and, thus, acid medium is required to supply the counter-ion A^- (Cl^-, HSO_4^-, $H_2PO_4^-$) for the solution doping of polyaniline. The concentration of dopant solution, doping time, and doping temperature can be varied to optimize the doping condition. The reversible nature of the solution doping process in the heterogeneous phase makes it less favorable, and hence a long time is required to achieve the optimum doping level for the conjugated polymer. A higher concentration of dopant could be one of the solutions to solve that problem, but in that case polymer degrades or over doping may result. In addition, room temperature solution doping is favored over heating as it can cause dedoping for the doped conjugated polymer. However, solution doping of conjugated polymer can be done under heating in special cases. As for example, poly(m-aminophenol) film was doped with sulfuric acid using the solution doping technique at a 100°C for 8–9 h to achieve the highest doping level [67]. The set-up used for the doping is shown in Figure 3.3. Heating only reduces the time for doping and helps to incorporate the anion to the polymer chain by breaking the intramolecular hydrogen bonding, as shown in Scheme 3.1. Polyaniline can be doped with corresponding anions by thermal acid–base exchange reaction method during film casting from N-methyl pyrrolidone solution using ammonium or alkylamine salt of protic acids p-toluene sulfonate, dodecyl benzene sulfonate, butyl naphthalene sulfonate, naphthalene sulfonate and camphor sulfonate [68].

In rare cases, the simultaneous film casting and doping can be done from the homogenous mixture solution of conjugated polymer with doping reagent in suitable solvent. The homogeneous mixture of silver hydroxide ammonium complex and poly(m-aminophenol) in dimethyl sulfoxide (DMSO) produced silver nanoparticle-doped polymer film at 140°C [55]. Free-standing film of doped polyaniline with Lewis acid $FeCl_3$ or $SnCl_4$ polymer were obtained by evaporation of soluble Lewis acid-polyaniline base complex from the nitromethane solvent in vacuum [69].

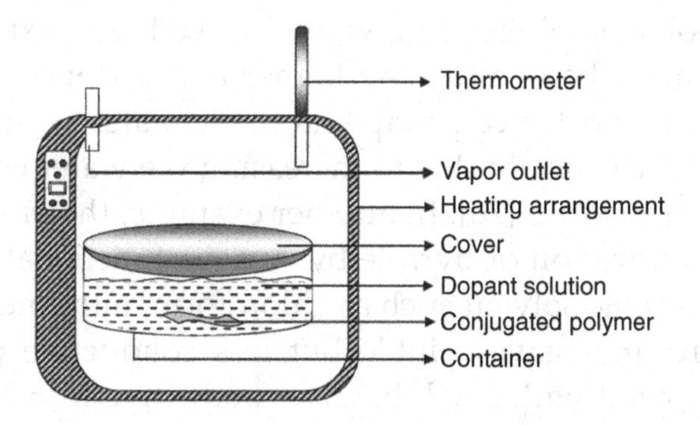

Figure 3.3 Solution doping set-up for conjugated polymer.

Scheme 3.1 Structure of poly(m-aminophenol) with intramolecular hydrogen bonding [54, 55, 67].

3.4 *In-situ* doping

In *in-situ* doping the polymers are synthesized directly in doped form as the polymer takes up dopant counter ions from the polymerization medium. *In-situ* doping occurs for the conjugated polymer during the chemical or electrochemical synthesis of the polymer from the corresponding monomer. The condition for this process is that the monomer and the doping agent must be soluble in the same medium or the monomer must be soluble after interacting with the doping agent. As the doping agent must be an electrolyte, the higher polarity of the solvents like water, water ethanol mixture, water DMF mixture and water DMSO mixture is favored. The *in-situ* HCl-doped polyaniline was synthesized in a binary solvent medium having different proportions of dimethyl formamide and water. It was observed that with an increase of water in the solvent mixture a tendency of precipitation of the polyemeraldine salt was observed. An explanation for this might be because the higher *in-situ* doped polymer association, and hence precipitation, facilitates with increasing dielectric constant due to increasing the water content of the solvent mixture [70]. In another example, the *in-situ* doping polymerization of pyrrole by various functional sulfonic acids in organic solvent such as chloroform, nitro methane or tetrahydrofuran gave soluble but less conductive polypyrrole, compared with insoluble but conductive (σ = 18 S/cm) or doped polypyrrole prepared in water [71]. This is because in less polar organic solvent the molecular weight of the synthesized polymer is much less and the doping interaction between the conjugated polymer and the dopant counter ion is weak. In the *in-situ* doping process, the doping agents first react to the monomer present in the polymerization medium and then polymerization of monomer associated with dopant ion occurs. As monomer has better reactivity than polymer, the *in-situ* doping process is a very effective method for strong interaction between the doping site and the doping agent. As for example, *in-situ* doped polyaniline and its derivatives can be chemically or electrochemically synthesized in various organic- or inorganic-acids medium where the acids are associated first with the monomer aniline base [72 ,73]. The concept can be well explained from the polymerization mechanism of aniline in HCl medium (Scheme 3.2). The electrochemical *in-situ* doping, which is already discussed in the section on electrochemical doping during polymerization, results in the doped conjugated polymer on the electrode. In *in-situ* doping, we can avoid the extra doping step as the synthesized polymer is simultaneously doped during the polymerization, and doped polymer pallet or film can be used for the application.

The main problem is that the *in-situ* doped polymer shows poor processability as it is not generally soluble in common organic solvents. In such cases *in-situ* doped conjugated polymers are dedoped first by the chemical method to make it soluble, and then after processing the polymer, is redoped to increase the conductivity. However, it would be convenient if the conjugated polymer salt could be doped *in situ* during its synthesis and made soluble in a solvent for direct film casting. It is reported that difficulties in the processing of polyaniline in the conducting form could be overcome by the use of functionalized protonic acid. A series of "functionalized protonic acid" such as $H^+(M^-R)$ in which the counter anionic species, (M^--R), contains a R functional group could be chosen to be compatible with nonpolar or weakly polar organic solvents [41]. Two examples are dodecyl benzene sulfonic acid or camphor sulfonic acid. These long alkyl chains on the phenyl ring lead to solubility in common organic solvents such as

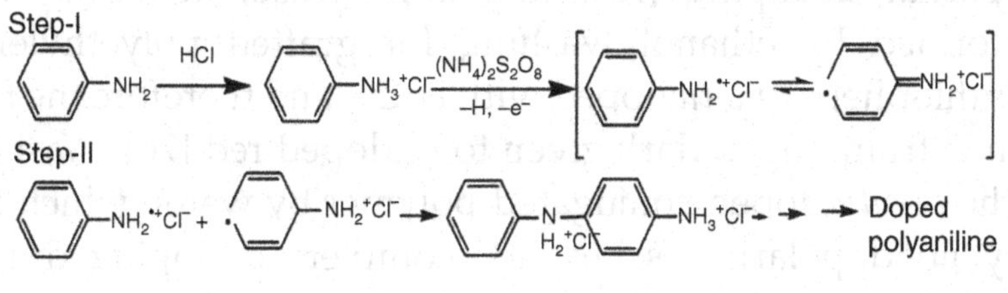

Scheme 3.2 Polymerization mechanism of aniline in HCl medium.

toluene, xylenes and chloroform, and the anionic part of the molecule (SO_3) dopes the polyaniline forming a complex that is conducting and soluble. The effects of functional sulfonic acids (e.g., 5-butylnaphthalene sulfonic acid, β-naphthalene sulfonic acid, *p*-dodecyl benzene sulfonic acid, camphor sulfonic acid, *p*-methylbenzene sulfonic acid, camphor sulfonic acid, *p*-methylbenzene sulfonic acid, *p*-hydroxyl benzene sulfonic acid, 5-sulfo-isophthalic acid, alizarin red acid and 8-hydroxy-7-iodo-5-quinoline sulfonic acid) on solubility and electrical and thermal properties, as well as morphology of polypyrrole prepared by *in-situ* doping polymerization of pyrrole in the presence of sulfonic acid as a dopant, have been investigated [71]. The dedoping may result on workup due to washing of *in-situ* doped precipitated polymer mass with suitable solvent for several times to remove the unreacted monomers, oligomers and byproducts. For example, the *in-situ* acid doped polyaniline thin film is deposited on silane-modified substrate blown-dried with nitrogen gas without washing, as washing with solvent might cause deprotonation [74]. The polarity of the washing solvent plays an important role, and as the polarity increases the dedoping becomes faster by better solubilizing the dopant electrolyte. Therefore, generally, less polar solvent like diethyl ether is preferred during the washing of *in-situ* acid doped polyaniline. However, some less polar solvent, which can effectively interact with dopant counter ion, may also cause partial or even complete dedoping for the *in-situ* doped conjugated polymer. As for example, the washing of *in-situ* doped 3,4-propylenedioxythiophenes with p-toluene sulfonate hexahydrate by isopropanol resulted in partial dedoping [75]. In another case, dedoping was performed by ethanol washing for grafted polyethylene–polythiophene *in-situ* doped with $FeCl_3$, and therefore the film turned from doped dark green to undoped red [76]. Washing of the *in-situ* doped conjugated polymer by water, which has very good polarity, is best for complete dedoping during workup. The dedoped form of polyaniline nanofibers *in situ* doped with HCl or camphor sulfonic acid was prepared by washing with water [77, 78].

3.5 Radiation-Induced Doping or Photo Doping

High energy radiation (e.g., γ-ray, electron-beam, neutron radiation, etc.) is used for radiation-induced doping or photo doping of the conjugated polymer. In this type of doping, when the undoped conjugated polymer is exposed to radiation of energy greater than its band-gap, electrons are promoted across the gap and the polymer undergoes "photo-doping." Alternatively, photo doping or radiation-induced doping can be explained as the separation of electrons and holes when a potential is applied by irradiation and, thus, photoconductivity is observed. Part of the conjugated polymer is locally oxidized, while the other part is reduced by photo absorption, and overall free charge carrier is observed due to charge separation within the matrix as follows [79]:

$$(\pi\text{-conjugated polymer})_n \longrightarrow [(\pi\text{-conjugated polymer})^{y+}(\pi\text{-conjugated polymer})^{y-}]_n$$

Thus, the lowest energy state or ground state of the conjugated polymer only recombines with proper symmetry due to photoexcitation. This recombination of ground state may result in either a radiative (luminescence) or nono-radiative one. The photodoping of some conjugated polymers, e.g., poly-p-phenylene, poly-p-phenylene vinylene, etc., exhibits high luminescence efficiencies, which is useful for the optical application of the polymer [79]. The photodoping scheme for polyacetylene is shown in Scheme 3.3. As for evidence, distinct spectroscopic characteristics of solitons can be observed in appropriate experimental conditions for the photo doping of polyacetylene. Generally the conjugated polymer doped

Photodoping

Scheme 3.3 Photodoping of polyacetylene.

by this technique shows only unstable photoconductivity, i.e., conductivity during photo exposure. As the doped charged center is not associated with any counter ion for stabilization, the photoconductivity for conjugated polymer only lasts until the excitations are either trapped or decay back to the ground state.

The doping of polythiophene has been studied *in situ* by utilizing the radiation-induced doping effect in an SF_6 atmosphere. From ESR characterization of the doped polymer it is confirmed that the carrier species for doping are polarons and the photodoped polythiophene is stabilized by the SF_5^- counter ion [80]. The effect of 9 to 250 Gy γ-radiation on the electrical conductivity of polyaniline in various oxidation states has been studied. The resistance of emeraldine base and leucoemeraldine salts decrease drastically, while the resistance of emeraldine salt slightly decreases due to a photo doping effect of the polymeric material induced by radiation [81]. X-ray radiation on thin films of chemically synthesized polyaniline and poly(o-methoxy aniline) improves the electrical conductivity under vacuum or dry oxygen atmosphere or humid atmosphere. Polyaniline or its derivative can be photo-induced doped by X-ray radiation for applications in lithography and microelectronics [82]. The solution cast films of polyaniline/poly(vinylidene chloride-co-vinyl acetate) and polyaniline/poly(vinylidene chloride-co-vinyl chloride) and polyaniline/poly(vinyl chloride) blends were exposed to 500 kGy gamma rays under ambient conditions. The increase in the conductivity of initially nonconducting films was highest for polyaniline/poly(vinyl chloride) blends reaching values of 10^{-2} S/cm from initial values of 10^{-7} S/cm [83]. The onium salt may be blended with the polyaniline to dope the polyaniline/onium system by electron-beam exposure. This is because onium salts decompose upon irradiation generating protonic acids, which act as an *in-situ* dopant for the polymer.

3.6 Charge Injection Doping

Charge injection doping through metal/organic interfaces has been employed. In this doping process the electrons are injected from the metal surface into the filled bonding π MO of the conjugated polymer and the polymer is oxidized [79];

$$(\pi\text{-conjugated polymer})_n - y(\text{Electrons}) \longrightarrow [(\pi\text{-conjugated polymer})^{y+}]_n$$

Otherwise, holes are injected from the metal surface into the empty anti-bonding π MO of the conjugated polymer and the polymer is reduced;

$$(\pi\text{-conjugated polymer})_n + y(\text{Electrons}) \longrightarrow [(\pi\text{-conjugated polymer})^{y-}]_n$$

The main difference between the charge injection doping and the chemical or electrochemical doping is that there is no counter ion associated with the charged center. So, this type of charge injection doping is very much unstable and the electrons or holes reside within the conjugated polymer only as long as the biasing voltage is applied. The transport of injected charges in metal/organic interface from the contact metal into the conjugated polymer channel causes the doping for the polymer by charge-transfer (Figure 3.4). The understanding of the formation of the metal–organic contact and the parameters which control the injection current should be helpful in understanding

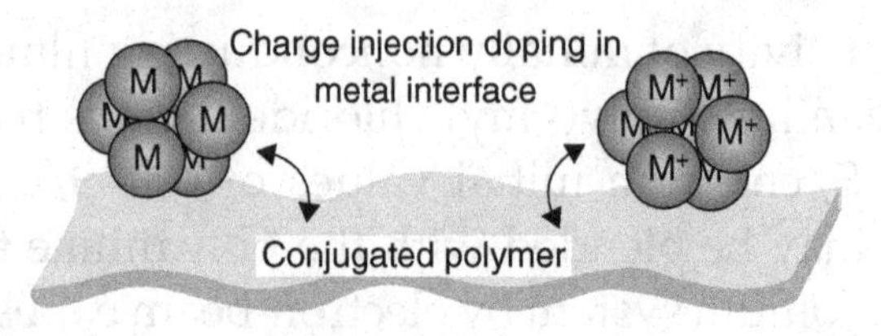

Figure 3.4 Charge injection doping of conjugated polymer in metal interface.

this type of doping mechanism [84]. Charge-injection doping is most common for metal/insulator/semiconductor (MIS) devices in which a metal and a conducting polymer are separated by a thin layer of an insulator. For those devices the charge injection occurs due to application of an appropriate potential across the structure by interface tunneling and hopping through charge trapping sites. The charge injection of poly(3-hexylthienylene) has been extensively studied on Au, Pt, etc., metal interfaces [85, 86].

4

Role of Dopant on the Conduction of Conjugated Polymer

4.1 Introduction

The parameters by which the phenomenon of electrical conduction is expressed are the mechanism of conduction by electrons and how the electron energy band structure of a material influences its ability to conduct. These principles are well extended to good conductors, semiconductors or insulators. The good conductivity of the metal can be explained from the consideration of the "electron sea" model. Metallic bonding is the electromagnetic interaction between delocalized electrons, which is called an "electron sea" around the metallic nuclei. So, it can be said that a metallic solid is a sort of ionic solid in which the free electron is donated to all the other atoms in the solid. Those loosely bound mobile electrons, which eventually form the conduction band, are responsible for good conductivity of the metal. The required energy to promote electron from valence band to conduction band is responsible for the conduction occuring and is known as "forbidden gap" or "band gap." The delocalized electrons can cross grain boundaries as they are free to move throughout the structure in three dimensions. Even though the pattern may be disrupted at the boundary, as long as atoms are touching each other, the metallic bond is still present. The conductivity (σ), which is reciprocal of resistivity (ρ) for metal can be measured using the empirical Ohm's Law:

$$V = IR \qquad \text{Where, } R = l/A$$

Where, V, I, R, l and A are voltage, current, probe distance and area respectively. The conducting metals, which are called Ohmic materials, fully obey Ohm's Law. But not all materials obey this law, e.g., one dimensional conductor (a linear polymer chain), gas discharge, vacuum tube, conventional semiconductors, etc. For an insulator, e.g., rubber, Bakelite, wood, etc., the promotion of electron to the conduction band from valence band is very difficult or not at all possible due to very high band gap energy. However, the materials having moderate forbidden gap have limited conductivity as the top empty states of the filled valence band are not available. Therefore, electrons must be promoted across the energy band gap into empty states at the bottom of the conduction band. To reduce the band gap some external force (doping) is needed by adding or withdrawing the electrons from the system. The extra added electron (electron donation) or created hole (electron withdrawing) in the lattice can move throughout the lattice of the inorganic semiconductor. Diffusion, by which the electron or hole can move from a region of higher concentration to one of lower concentration inside the lattice, describes the net charge transport of molecules. The diffusive descriptions are essential for the understanding of dopant movements and can be confirmed by the suggested specific successful models which apply Fick's Law and boundary conditions to derive it. For these semiconductor materials the conductivity depends on the number density (number of electron; n) of charge (e is the electron charge) and how fast they can move in the materials (charge mobility; μ) throughout the lattice by diffusion mechanism. The conductivity of such materials may be expressed as [87]:

$$\sigma = n\,|e|\mu$$

The measurement of conductivity for the conjugated doped polymer is similar with the semiconducting materials. However, the conduction mechanisms for conjugated polymers are slightly different than the conventional inorganic semiconductors. The interaction of dopant molecules can play an important role in the conduction process of doped conjugated polymer, as doping enhances the conductivity of conjugated polymer by several fold. Due to doping interaction, charge centers are generated within the conjugated polymer and the counter ion is associated with the charged center in the polymer matrix to maintain the electroneutrality. The conduction in doped conjugated polymer results in the generation and disappearance of charged sites on the conjugated polymer chains, and electroneutrality of the doped conjugated polymer is maintained by the incorporation and repulsion of mobile counter ions. The interaction of dopants in conjugated polymer is different than that in inorganic semiconductor, as very high dopant concentration is required for the doping of conjugated polymer. In fact, the conduction efficiency should have a threshold at a certain dopant concentration, with low conduction below that concentration. Thus, the conduction of doped conjugated polymer is thereby usually assumed to be dependent on the following factors:

i. Charge defects within doped conjugated polymer
ii. Charge transport within the doped conjugated polymer
iii. Migration of dopant counter ions

4.2 Charge Defects within Doped Conjugated Polymer

Doping is a reversible process that occurs readily by interchanging the redox state of polymer without any degradation or major change. Due to doping, i.e., the addition of a donor or acceptor species to the conjugated polymer, the defects are incorporated into the polymer matrix. The names of the species having such defects in the conjugated polymer matrix are soliton, polaron and bipolaron. The charged soliton, polaron and bipolaron structures of polyacetylene and polypyrrole are shown in Scheme 4.1. The doping affects the band gap energy of the undoped conjugated polymer. The orbital structure as well as band structure for undoped and doped polymer is shown in Figure 4.1. As a result of doping, energy of the HOMO state increases or that of LUMO decreases and thus increases in conductivity.

4.2.1 Soliton

In the doping process, single electron donation can occur to the polymer (Pn) by the dopant (donor) or single electron acceptance can take place by the dopant from the polymer (acceptor); the resultant species is known as "neutral soliton."

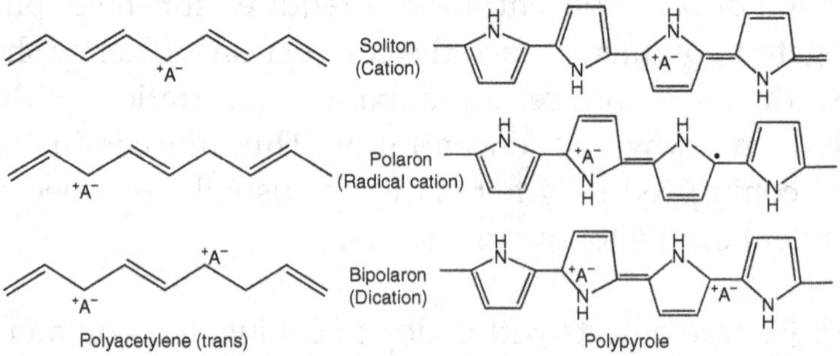

Soliton (Cation)
Polaron (Radical cation)
Bipolaron (Dication)
Polyacetylene (trans)
Polypyrole

Scheme 4.1 Soliton, polaron and bipolaron for trans-polyacetylene and polypyrrole.

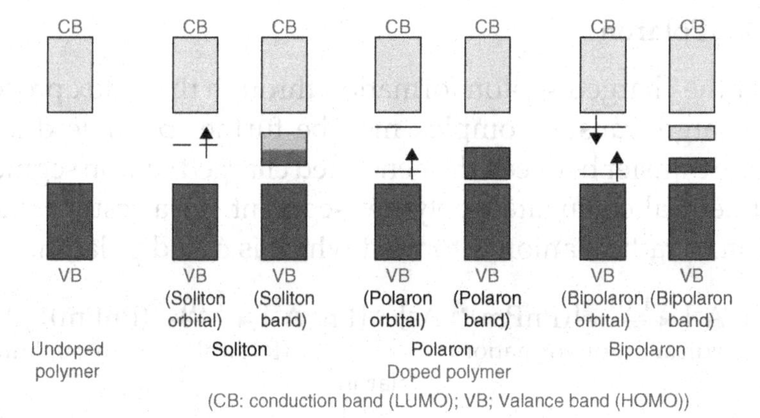

Figure 4.1 The orbital and band structures for undoped and doped (soliton, polaron, bipolaron) conjugated polymer.

$$\text{Pn} \xleftrightarrow{\text{A}} [\text{Pn}^{\bullet}\ \text{A}^{\bullet}]$$

Neutral soliton

If the pair of electrons are accepted (cation) or donated (anion) by the dopant then charged soliton is formed.

$$\text{Pn} \xleftrightarrow{\text{A}} [\text{Pn}^{+}\ \text{A}^{-}] \qquad \text{Pn} \xleftrightarrow{\text{A}} [\text{Pn}^{-}\ \text{A}^{+}]$$

(Reduction) (Oxidation) (Oxidation) (Reduction)

Charged soliton

In soliton the independent charge defects are observed due to two-phase separation of bonding electron to an opposite orientation having an identical energy state. Thus, the soliton state is a type of degenerate ground state, which is introduced due to oxidation or reduction, i.e., doping of the conjugated polymer. In the degenerate ground state the corresponding electrons are not bound with high energy binding and thus can freely spread along the conjugated chain. As for example, the solitons in doped polyacetylene are observed to be delocalized over about 12 CH units [87]. This soliton state results in soliton band, which appears in the middle of the band gap as a new localized electronic state to conduct electricity.

4.2.2 Polaron

After the charged soliton formation through the redox process the charge transfer complex may be further possible due to charge transfer between the generated charged soliton segment and neutral conjugated polymer segment. As a result radical cation or radical anion is formed which is called polaron.

$$[Pn^{+} A^{-}] \overset{Pm}{\longleftrightarrow} [(PnPm)^{\bullet +} A^{-}] \quad [Pn^{-} A^{+}] \overset{Pm}{\longleftrightarrow} [(PnPm)^{\bullet -} A^{+}]$$

(Radical cation formation) (Radical anion formation)

Polaron

where, Pn, Pm denote polymer segments. Thus the polaron, which is either a radical cation or a radical anion, is the combination of a charge site and a radical. This polaron state results in a new localized electronic state close to the lower energy state, which contains an unpaired single electron. The energy level associated with the polaron represents a destabilized bonding orbital and thus has a higher energy than that of the valence band. For doped polypyrrole, the polaron states are about 0.5 eV from the lower valence band edges [9].

4.2.3 Bipolaron

Instead of single electron transfer from charged soliton or polaron generation, the charged soliton formation may be followed by a couple of electron transfers with the dopant. This process results in the formation of a dication or dianion, which is known as bipolaron.

$$[Pn^{+} A^{-}] \overset{A}{\longleftrightarrow} [Pn^{2+} 2A^{-}] \quad [Pn^{-} A^{+}] \overset{A}{\longleftrightarrow} [Pn^{2-} 2A^{+}]$$

(Dication formation) (Dianion formation)

Bipolaron

So, the bipolaron has new spinless defects and it is created by further oxidation of polaron. The bipolaron state can be considered as a combination of two distinct polaron states, but it has a somewhat lower energy state than two polaron states. For bipolaron two new localized electronic states are located within the band gap containing an unpaired single electron by each. The bipolarons are about 0.75 eV for doped polypyrrole [9]. The two bipolaron states within the band gap can overlap for heavily doped conjugated polymer to form a bipolaron band, which acts as a partially filled band to show high conduction.

4.3 Charge Transport within the Doped Conjugated Polymer

The doped conjugated polymers possess charged species like soliton, polaron and bipolaron. The charged centers are mobile and can move along the conjugated polymeric chain by the rearrangement of the double and single bonds in the conjugated system that occurs in an electric field. Conduction by polarons and bipolarons is the dominant mechanism of charge transport in polymers with nondegenerate ground states. The movement of these charge carriers is the main mechanism for the conduction of electricity by the doped conjugated polymer. In the case of undoped conjugated polymer the lack of charge center as well as carrier site is the reason for its very low conduction. As already discussed in previous sections the amount of dopant used in the case of conjugated polymer is much higher than that of inorganic semiconductors. So, the number of charges or concentration of charge species in the case of conjugated polymer is much higher than that in the inorganic semiconductors. The main difference for the electrical conduction in doped conjugated polymer is that the diffusion mechanism for charge transport like doped inorganic semiconductor does not fit well. The doped conjugated polymer matrix has very low electrical permittivity, which means the charge centers in the polymer chain must be bound to the parental dopant counter ion by a strong coulombic binding energy. The charge transport inside the doped conjugated polymer is the combination of the effects of large coulomb correlation between charge carriers and strong polaronic coupling between localized hopping states. Thus the key feature for the charge transport in doped conjugated polymer is that it is taking place by hopping in a distribution of more or less localized states according to the following electronic parameters.

4.3.1 Electronic Parameter Responsible for Charge Transport

According to the equation for conductivity of semiconductor material, the conductivity of conjugated polymer should be directly related to the concentration and mobility of the charged species inside the polymer, i.e., electronic structure of the doped conjugated polymer. The electronic structure of conjugated polymers depends on a number of electronic parameters such as ionization potential (IP), electron affinity (EA), band gap (E_g) and bandwidth (BW). The relationship of these parameters to the polymer band structure is illustrated in Figure 4.2.

The IP and EA indicate the relative ease of ionization, whereas the band gap and bandwidth controls the interchain transport within ionized conjugated polymer material to conduct electricity. IP can be defined as the energy required for removing electron from the conjugated polymer into the vacuum, and it is the fundamental measure of aptitude for acceptor doping in conjugated polymer. Alternatively, EA reflects the relative case of addition of an electron from the vacuum to the bottom of the higher energy band in a conjugated polymer. The EA is the measure of aptitude for donor doping in conjugated polymer. A small value for the IP indicates a polymer which is easily oxidized, and a large value for EA indicates a polymer which is easily reduced. IP or EA are only responsible for the generation and stability of charge within the doped conjugated polymer. The factor limiting the conductivity is not only the carrier or charge concentrations,

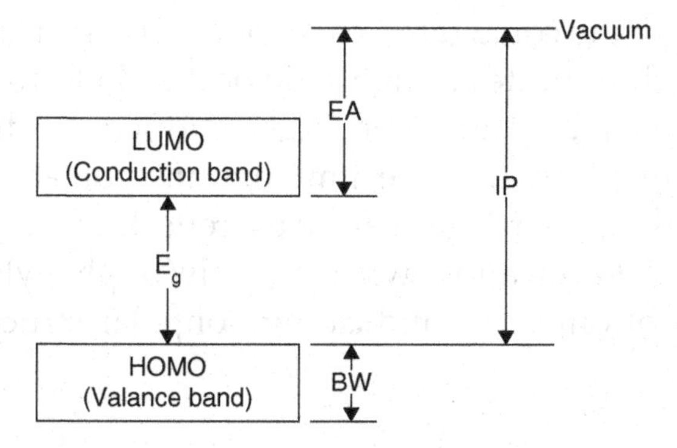

Figure 4.2 Relationship of π-electron band structure for conjugated polymer to vacuum and various energetic parameters. Eg is the optical band gap; BW is band width of the fully occupied valence band; EA is the electron affinity and IP is the ionization potential.

but also the carrier mobility. Many organic (and inorganic) syntheses are known that possess free radical or charged species, yet remain insulators. Thus the necessary condition for a polymeric system to be conductive is to possess both charge carriers and orbital system, which allow them to be mobile. The precise mechanism of conduction through the charge carriers like solitons and bipolarons is not yet fully understood. The problem is in tracing the path of the charge carriers through the doped conjugated polymers. This can be correlated to the diffusion of charge or hole within the conventional semiconductor materials. The influence of diffusion of the dopant ions within doped conjugated polymer on the overall behavior of the polymers has already been explained. Thus the essential condition for electronic conduction in conjugated polymer is that it must contain an overlapping set of molecular orbitals to provide reasonable carrier mobility along the polymer chain. Now, the band gap (E_g) is the difference between the IP and EA and correlates with the optical absorption threshold. In addition, BW is the width of a particular band and it is correlated to the carrier mobility. For a conjugated polymer BW is usually the highest occupied band of states with π-electron character. Strong interactions among the π-electrons of the conjugated backbone result in a large bandwidth, which indicates a highly delocalized electronic structure. A large bandwidth favors high conductivity due to high intrachain mobility. As for example, trans-polyacetylene has a bandwidth in the range of 6–10 eV reflecting a high degree of electron delocalization, whereas, poly(p- phenylene) has a bandwidth of only 1 eV indicating nonpolar structure in its backbone.

4.3.2 Charge Transport Mechanism

The ambipolar transport mechanism, i.e., the ability to transport both electrons and holes, has now been well accepted for several doped conjugated polymers. In a favorable electronic environment, the charge transport inside a doped conjugated polymer can be considered as simply a better delocalization of electrons and holes throughout the π-conjugated backbone. The process is a type of intrachain mobility of the charge center associated with the dopant counter ion. In general, intrachain mobility is not the only limiting factor in determining the conductivity of a doped conjugated polymer. This is because the π-conjugation is disturbed, as all conjugated polymers considered to be highly disordered have crystalline and amorphous regions. In the absence of any external potential, transport is purely diffusive and in general the charge transport through diffusion is possible in all directions. Hence, in doped conjugated polymers there are three directions for the carrier mobility through diffusion inside the polymer network [87]:

a. Intrachain or intramolecular transport (delocalization)
b. Interchain or intermolecular transport (hopping)
c. Interparticle transport (percolation)

These three processes determine the effective mobility of the carriers for doped conjugated polymer, a complicated resistive network (Figure 4.3). Thus the mobility, and therefore the conductivity, are determined on both microscopic (intra- or interchain) and a macroscopic (interparticle) level. Although the mechanisms responsible for interchain transport are not well understood, there are a number of factors which will clearly affect carrier mobility between chains. Structural factors in doped complex such as chain planarity, degree of crystallinity, chain kink and crosslinks all play a role in interchain transport. In addition, electronic coupling between chains can be affected by the dopant ions, where both its size and shape can control the structure of dopant polymer assay. The coupling may be further enhanced by appropriate orbital overlap between dopant molecules and polymer chains.

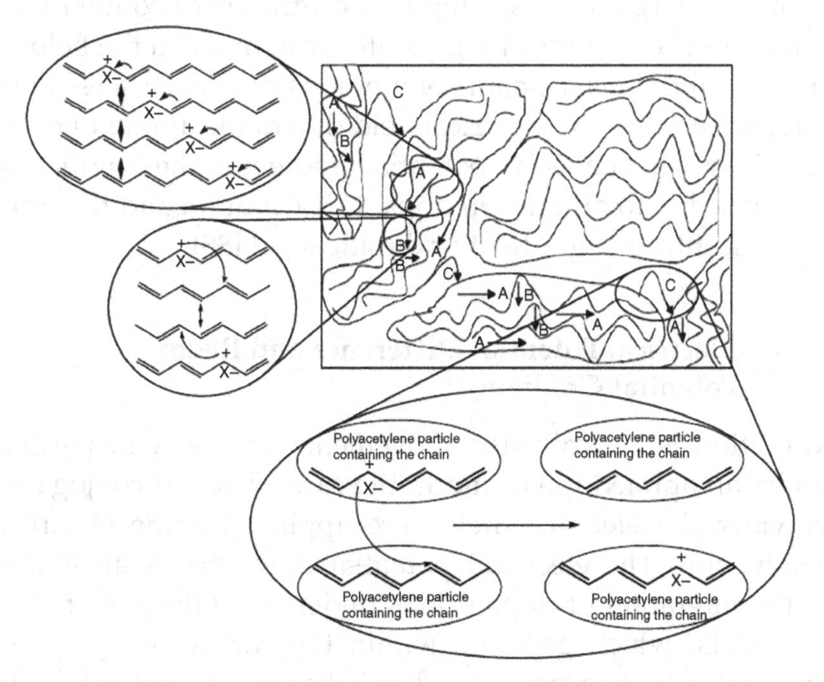

Figure 4.3 Conductivity network of a conducting polymer with (A) indicating intrachain transport of charge, (B) indicating interchain transport, (C) indicating interparticle transport and arrows showing path of charge carrier migrating through the material.

4.4 Migration of Dopant Counter Ions

As a result of the doping process, a counter ion is incorporated into the charged conjugated polymer matrix and, thus, the complex is electrically neutral and conducting. The conduction in doped conjugated polymer requires transport of the counter ions with the charge centers within the polymeric network to compensate the charges. So, the conduction within a doped conjugated polymer can be considered as migration of the charge-neutral states. The basic difference between the electrical properties of doped conjugated polymers and those of inorganic semiconductors is the involvement of both charge centers as well as counter ions in a conduction mechanism for the doped conjugated polymers. Since the counter ions within the doped conjugated polymer matrix are not very mobile with the charge centers, a high concentration of counter ions is required so that the charge centers can move in the field of close counter ions. Dopant counter ions migrate as a result of three influences: the electrical field due to the applied potential difference, the redox-potential gradient in the conducting polymer due to an uneven dopant distribution and the dopant concentration gradient or doping level [88].

4.4.1 Electrical Potential Difference and Redox-Potential Gradient

Migration is basically the movement of a charged particle under an applied force, and in the case of doped conjugated polymer the electrical fields are applied for the electrical conduction. The electrical potential difference is attributed to the migration of dopant ions under the influence of electric fields, which generate within the conjugated polymer due to the applied potential. The behavior of conductivity in doped conjugated polymer can be explained on the basis of the existence of potential barriers between highly conducting regions like inorganic semiconductors. These barriers are due to conjugational defects or other inhomogeneities in the polymer chains. The charge carriers will have to hop or tunnel through the potential barriers. The doping creates charged centers like solitons, polarons and bipolarons with dopant counter ions and during hopping of those charged centers from one particle to another, will create a potential difference between the two particles. These charged centers lead to a break in bond alternation, i.e., the phase is opposite on the two sides of the charge center. Partial delocalization occurs over several monomer units, and the units deform structurally. Thus, the application of electric field results in charge carrier motion from a trapping center. The created charged centers and the associated dopant counter ions are in equilibrium at the applied potential difference, and the system must allow sufficient time to fully adjust to the new charge distribution. In the conjugated polymer, a semicrystalline material of ordered crystallites is interconnected by amorphous regions. It is well known that the packing order of the conjugated polymer chains, and hence the distance between two adjacent chains, changes upon doping. For conduction inside the doped conjugated polymer the current must pass through the above disordered regions. It has been observed that doped conjugated polymers having the most homogeneous chain structures appear to be the best conductors. On the other hand, heterogeneity can yield carrier localization on the conjugate chain, which provides the lowest potential for the charged center. Generally, the above migrations of dopant ions are reversible in nature, i.e., the dopants migration occurs in the other direction after reversing the potential difference across the conducting polymer.

Under an applied electric field, the counter ions migrate towards the oppositely charged electrode. At the same time the neutrality is maintained within the doped conjugated polymer by diffusion of the counter ions with the charged center due to coulombic attraction. These two simultaneous processes are contradictory in nature. The redox potential gradient builds up due to simultaneous migration and diffusion of dopant counter ion in the opposite direction. The migration of charge center increases the overall potential, as the process is faster than that of the counter ion diffusion. At very low applied potential difference the doped conjugated polymers behave like Ohmic materials, that means they obey Ohm's law. This is because in very low applied potential difference the migrations of dopant counter ions become negligible and electrical conduction occurs only due to the charge transport with the diffusion of the counter ions. The redox potential gradient increases gradually with the increase of applied potential difference, and hence the charge separation of counter ions from charge centers within the doped polymer matrix. Due to this reason the Ohmic behavior disappears, or even loss of conducting property is observed, when the doped conjugated polymer is exposed to very high potential difference for a long time.

4.4.2 Dopant Concentration Gradient or Doping Level

From the previous section it can be seen that the electrical conductivity of doped conjugated polymer depends on two important factors, the number of carriers (e- or holes) and charge carrier mobility. Due to doping the charged species like soliton, polaron and bipolaron are incorporated into the polymer matrix, and increasing the doping percentage will increase the density of charge carriers. So, in that sense, the doping percentage will gradually increase the conductivity of the doped conjugated polymer. But this is not the case, as relative mobility of those charged species also controls the bulk electrical conductivity of the doped conjugated polymer.

Increase in the charge density due to doping in conjugated polymer provides more electrons in the conduction band, but not much of an increase or even a fall in the conductivity is observed due to lower conjugation length after a certain concentration of dopant. That means the distance between charged centers within the highly doped conjugated polymer become closer. For these closer distances of charged species, the mobility of those species are restricted within the polymer. This happens above a certain optimum concentration of dopant for a particular conjugated polymer and can be termed as "doping level." For conjugated polymer the dopant concentration is given in mol %, taking one polymer doped center as a mole. As for example, for iodine-doped polyacetylene the doping level can be described as the mol % of iodine taking one CH unit as a mole [89]. It is observed that upon increasing the doping level the conductivity rises by many orders of magnitude, but there is no sharp transition from an insulating to a metallic state. Very light doping, far below 0.1%, has a drastic effect on the conductivity. The optimum doping level for almost all the conjugated polymer is 0.5 mol % and it is in that concentration of dopant that the conductivity of doped conjugated polymer is generally highest. Going to a higher doping level, e.g., > 0.5%, results in charge defects in every second monomeric subunit. Therefore, the conductivity of the doped conjugated polymer beyond optimum doping level remains unchanged as mobility of charges restricts the coulombic repulsion and is explained as an over-oxidation effect [90]. The simulated plot for conductivity of conjugated polymers as a function of doping level can be consider as a standard plot to compare with measured conductivities of the doped conjugated polymers [88]. This effect is well reported for doping of conjugated polymers like polyacetylene, polyaniline and poly(m-aminophenol) (Table 4.1). Initially little attention was paid to the fact that a shift of the Fermi level, which is of utmost importance for electrical properties, should be expected on varying the doping level. Generally, for the performance of doped conjugated polymer in electronic devices, the knowledge of the Fermi level as a function of the doping level is a necessary condition for the utilization of these materials. By adjusting the doping level a wide range of conductivity can be obtained for a doped conjugated polymer anywhere

Table 4.1 Effect on conductivity with the doping level of various conjugated polymers.

Conjugated polymer	Dopant counter ion	Doping level with respect to counter ion (mol%)	Conductivity (S/cm)	Ref.
Poly(3-methylthiophene)	$CF_3SO_3^-$	0.3	30–50	25
		0.5	100	
Polyacetylene	ClO_4^-	0.1	1000	23
		0.12	50	
Polyacetylene	Li^+	0.1	10–100	23
		0.17	200	
		0.2	200	
Polyacetylene	K^+	0.16	50	23
		0.17	500	
Polyaniline	Cl^-	0.34	0.03	91
		0.38	0.036	
		0.52	0.1	
		0.60	0.03	
		0.75	0.02	
Poly(m-aminophenol)	Na^+	0.51	10×10^{-6}	54
		0.53	0.85×10^{-6}	
		0.62	23.4×10^{-6}	
		0.70	6.6×10^{-6}	
		0.58	5.46×10^{-6}	

within the insulating or semiconducting (non-doped) and highly conducting region (optimum doped). Good evidence of doping can be explained by measuring the concentration of dopant in the doped conjugated polymer. The methods that can be used to determine the weight uptake by the conjugated polymer are the simple gravimetric method, elemental microanalysis method, Schöniger combustion method, Micro-Kjeldal elemental analysis method, flame photometric method, etc. But a very small change of mass due to doping in conjugated polymer is preferably and precisely determined by the elemental microanalysis method. This method is a very popular method and is used to determine the dopant concentration in most studies. The Schöniger combustion method is used to determine the Cl percentage in the sample along with the elemental microanalysis method to get the doping level of HCl-doped polyaniline [91]. But a restriction of the elemental microanalysis method is that only a few elements (C, N, S, O) can be detected by this method. To overcome this, the Micro-Kjeldal elemental analysis method can be used. Another method, the flame photometric method, can be used very effectively to determine the percentage of metal dopant. As for example, the percentage of sodium ion doping level in poly(m-aminophenol) is determined by combining Micro-Kjeldal elemental analysis and the flame photometric method [54]. The energy-dispersive X-ray spectra (EDS) method combined with the elemental microanalysis method is also used to determine the doping level of poly(m-aminophenol) by inorganic acids [92]. Electrochemical quartz crystal microbalance (EQCM) measurements were also used during the n-doping of poly(3,4-ethylenedioxythiophene) films exposed to a range of electrolytes tetraethylammonium tetrafluoroborate, tetrapropylammonium tetrafluoroborate, or tetraethylammonium tosylate to determine the dopant concentration inside the polymer [93].

5

Influence of Properties of Conjugated Polymer on Doping

5.1 Introduction

The novel unique properties of conducting polymer that are not typically available in other materials are generally related to the fundamental electronic structural feature of the polymer. The doping, i.e., creation of a certain amount of positive or negative charge within the conjugated polymer, results in a modification of their electronic structure, which also affects the properties of the doped polymer. The doping in conjugated polymer imparts dramatic changes in the electronic, electrical, magnetic, optical, and structural properties of the polymer. These novel properties of doped conjugated polymer enable a number of applications. Thus, the correlation of the electronic structure of doped conjugated polymer with the properties is an important ingredient in the applications. For example, the investigation of induced-doping interaction between conjugated polymer and various metals has become an important issue for optical device application. In view of this, the influence of properties of conjugated polymer on doping is discussed briefly in this chapter. The doped conducting polymer has several fold higher conductivity than that of the so called undoped form. This should be important evidence for doping. But the saturated insulating polymer (e.g., polyethyleneoxide) containing a high percentage of ionic salts (e.g., Li salts) can have electrical conductivity in a semiconducting region. Generally the doped conjugated polymer contains a high percentage of ionic species like soliton, polaron and bipolaron with the counter ions. That means the conductivity which is observed for the doped conjugated polymer may be misunderstood as ion conduction inside the polymer matrix. So, the change of properties of the doped conjugated polymer can also be considered as evidence to explain the doping phenomena in the conjugated polymer. Details are given below.

5.2 Conducting Property

The prime importance for doping in conjugated polymer is to improve the conductivity of the polymer. The role of doping on conduction of conjugated polymer has already been discussed in Chapter 4. The DC electrical properties of doped conjugated polymers have been extensively studied in recent years. In general, the I-V characteristic of the doped conjugated polymer by measuring the voltage (V) with varying current (I) at room temperature has been used to explain the electrochemical activity of the materials. A linear relationship of the I-V plot explains the Ohmic nature of the doped conjugated polymer comparable with that of metal. However, the linear nature of the plot is only valid for a short region of applied current but not within a broad range. It has been discovered that the conductivity of polyacetylene films can be increased from 10^{-9} to 10^{3} S/cm by doping and that the conductivity is a function of the degree of oxidation [20]. Pressed pellet of poly(p-phenylene vinylene) shows a conductivity 10^{-10} S/cm at ambient temperature, while after exposing the pellet to AsF_5 vapor the doped poly(p-phenylene vinylene) gives conductivity as high as 3 S/cm [94]. The conductivity of some important doped conjugated polymers achieved by researchers by varying the dopant is shown in Table 5.1. Data may also be available reporting on the maximum conductivity of doped conjugated polymer higher than this. However, from the table it is clear that the conductivity value for the conjugated polymer is not a constant or reproducible value even when the dopant remains the same at its optimum doping level. This is due to various reasons and instability of the dopant inside the polymer matrix is important among them. The plot σ/σ_0 (where σ is the conductivity with the time interval and σ_0 is initial conductivity of the doped conjugated polymer) vs. time can be considered in explaining the dopant stability inside the polymer. The parallel characteristic of plot σ/σ_0 vs. time with x-axis, as explained for sodium ion doped poly(m-aminophenol) in Figure 5.1, should indicate that the doping inside the polymer backbone is very much stable [54]. The dopant stability inside the conjugated polymer is the most important barrier for the real application of doped conjugated polymer instead of inorganic semiconductor.

Table 5.1 The conductivity of some important doped conjugated polymer (data extracted from 24,28).

Polymer	Conductivity region (S/cm)	Highest conductivity achieved (S/cm)
Polyacetylene	100–200	100000
Poly(p-phenylene)	1–500	1000
Poly(p-phenylene sulfide)	3–300	20
Poly(p-phenylene vinylene)	1–1000	1000
Polypyrrole	40–200	2000
Polythiophene	10–100	200
Polyaniline	10^{-2}–5	10

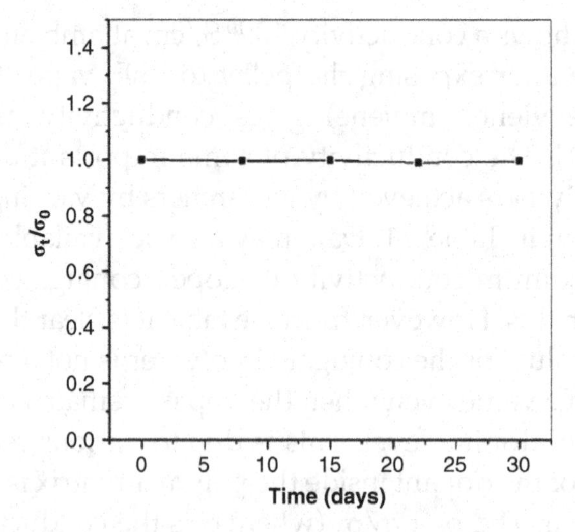

Figure 5.1 Change of DC-conductivity with time for sodium-ion-induced doped poly(m-aminophenol). Reproduced with permission from ref. [54], Copyright © 2010 Elsevier B.V.

The DC electrical properties of various doped conjugated polymers have been extensively studied in recent years, but relatively little work has been done on their AC electrical behavior. Recently, AC conductivity measurements were used as a novel tool for separating different conduction mechanisms in conducting polymers. Frequency-dependent AC-conductivity is very useful for the dielectric property study of conjugated polymer, especially for its composites. The frequency-dependent dielectric properties and electrical conductivity of some conducting polypyrrole (PPy) composite films have been studied for qualitative description of the conduction mechanism [95].

5.3 Spectroscopic Property

The spectroscopic properties of conjugated polymer refer to some important concepts proposed by solid-state physicists. Due to doping, electronic defects like soliton, polaron or bipolaron are formed and new energy levels are observed within the HOMO-LUMO level of conjugated polymer. On the basis of these new energy levels, the numbers of intra-gap transitions are also changed and the conjugation throughout the π-system increases. Thus, spectroscopic studies of electronically excited states are useful in order to have an idea about the change of properties of conducting polymers on doping. Common spectroscopic evidences, e.g., ultraviolet-visible (UV-Vis), Fortier transform infrared (FTIR) and Nuclear Magnetic Resonance (NMR) spectroscopy are the simplest, precise evidence for doping confirmation in conjugated polymer. Other uncommon spectroscopic evidence for doping of the conjugated polymer is also discussed here.

5.3.1 UV-VIS Spectroscopy (Optical Property)

The undoped conjugated polymer itself is often highly colored as it absorbs energy equal to the π - π* energy gap. The transmitted wavelength resulting in the observed color falls within the visible region. The red shift (absorption band shifting towards higher wavelength or red region) is the general observation for π - π* transition of doped conjugate polymer due to extended conjugation through doping. Depending on the nature of dopant and doping level, in the doped conjugated polymer the extra energy state or energy band is generated due to doping in between the valence band and conduction band. Thus, beside the π - π* transition, now the transitions may take place from the valence band to the lowest doping energy level as well as to the highest doping energy level. Furthermore, a third optical absorption is possible for transition between the two doping energy levels. The intense color change is observed due to these inter-gap transitions passing from the pristine to the doped state of the conjugated polymer. In the case of most doped conjugated polymers, these effects are responsible for a new intense peak at a higher region than the π - π* transition. Optical absorption spectra as a function of dopant concentration for polythiophene doped with ClO_4^- is shown in Figure 5.2 along with consideration of their different possible band transition [80]. The same observation is reported for poly(m-aminophenol) doped with various inorganic acids like sulfuric acid, perchloric acid and

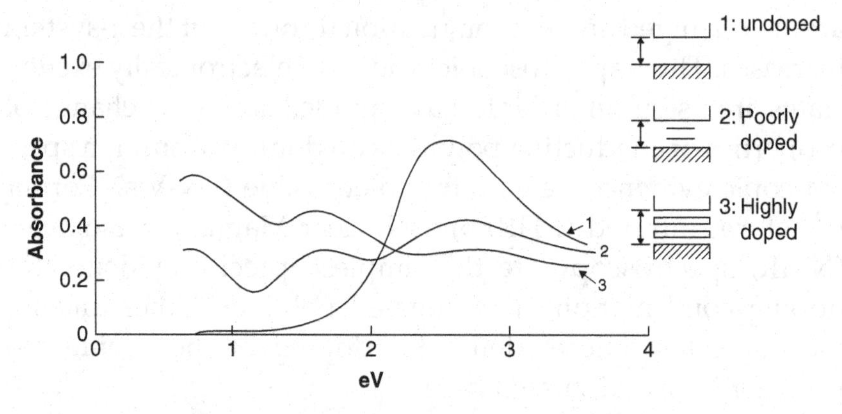

Figure 5.2 Optical absorption spectra with band structure of ClO_4^- doped poiythiophene as a function of dopant concentration. Reproduced with permission from ref. [96], Copyright © 1987 Pergamon Journals Ltd.

phosphoric acid [92]. In that communication, it is explained that the $\pi - \pi*$ absorption is red shifted to 400 nm from 350 nm for sulfuric acid-doped poly(m-aminophenol), and due to the polaron electron transition another new absorption maximum appears at 650 nm. While in the case of perchloric and phosphoric acid doping the polaron very weak absorption maxima appears at around 500 nm. The better doping effect of sulfuric acid in comparison to perchloric and phosphoric acids is the reason for such types of UV-VIS spectra.

5.3.2 FTIR Spectroscopy

The peak shift in FTIR spectra can also be observed for doped polymer compared to undoped polymer. Due to extended electronic conjugation, which lowers the bond strength in the doped polymer more than that of the undoped one, the position of the corresponding vibrational (FTIR spectra) peak shifts towards a lower wave number (bathochromic shift). The structural differences can be observed in the vibrational spectra based on changes in the position and shape of the characteristic FTIR bands for various bonds like N-H, C-H (aromatic), etc., of polyaniline doped with strong acids (HBF_4, $HClO_4$, HCl, H_2SO_4, H_3PO_4) or weak acids (formic, acetic acid, monochloroacetic acid, dichloroacetic acid, trichloroacetic acid) [97]. A shift in the C-H aromatic bands is observed for the acid-doped polyaniline depending on the doping ability of the acid due to changes in electron density and resonance. The broader, higher intensity peaks and a shift to higher energy is also consistent with an increase in hydrogen bonding with the N-H group for HBF_4 and $HClO_4$ dopant compared to phosphate, chloride and sulfate which have weak hydrogen bonding. The proton-doping mechanism in perchlorate-doped poly-3-methylpyrrole films at different oxidation states as well as after immersion in acid (pH=1) and basic (pH=12.6) aqueous solutions has been documented from the comparative study of infrared spectra spectroscopy [98]. The dopant interacts with the special group inside the conjugated polymer and can reduce the intensity of FTIR spectra of that particular group. The intensity of absorption peaks at 1639 and 1570 cm^{-1}, which are assigned to the C=N stretching of the quinoid and the C=C stretching of the benzenoid rings respectively, are significantly reduced in the case of polyaniline-carboxylic-acid functionalized carbon nanotube nanocomposites compared to those of pure polyaniline (Figure 5.3). According to the authors, the doping interaction between the quinoid ring of the polyaniline and the carboxylic-acid-functionalized carbon nanotube could be

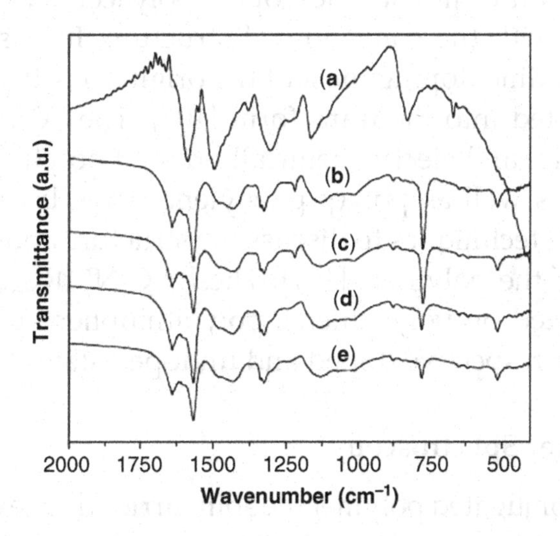

Figure 5.3 FTIR spectra of (a) pure polyaniline and (b) 0.5 % (c) 1 % (d) 2 % (e) 3 % carboxylic acid functionalized carbon nanotube-polyaniline nanocomposites. Reproduced with permission from ref. [99], Copyright © 2011 Elsevier B.V.

the result of the decrease in intensity of the peak [99]. The dopant having the special group which is already present with the conjugated polymer can enhance the intensity of FTIR spectra of that particular group. However, it is unnecessary to site specific examples for this evidence, as it can be found in most reports as a routine test that shows evidence of doping in conjugated polymer.

5.3.3 NMR Spectroscopy

The study of dopant interaction with the conjugated polymer by Nuclear Magnetic Resonance (NMR) is restricted, as the doped form of the conjugated polymer generally is insoluble in common organic solvent. One solution for this problem which is adopted in most studies is solid-state NMR. Solid-state NMR is very sensitive to the microscopic details of charge and spin distributions and dynamics. The slight shifts in NMR spectra (especially ^{13}C NMR) have occasionally been observed in the doped state of conjugated polymer due to the failure of hyperfine interaction of the nuclei with the polarons. The second moment of the continuous wave NMR spectra of pure and iodine-doped polyacetylene has been compared with the conventional structure. It was observed that with iodine doping most of the original cis-type of material converted into the trans form [100]. The ^{13}C-NMR spectra of neutral and electrochemically-doped poly(p-phenylene vinylene) as well as poly(p-phenylene) have been recorded by different techniques to discuss the structure and molecular mobility of the polymers [101]. These ^{13}C NMR studies have been reported on polyazulene, polybithiophene and polyfuran in their respective doped and undoped states [102].

5.3.4 Other Spectroscopy

In doped conjugated polymer the spin carrier charge can be well documented by spin dynamic characterizations like NMR or electron spin resonance (ESR). These methods have additional advantages to discuss such as the mechanism of anisotropic and high electrical conductivity in doped conjugated polymers. A reference is available which details NMR and ESR results for common conducting polymers like polyacetylene, polyaniline and polythiophene [103]. X-ray photo spectroscopy (XPS) has the capability to obtain information about both the chemical composition as well as the oxidation state, and so it has been utilized as an analytical tool to determine the doping properties of the conducting polymers. Sulfonated polyaniline has been studied using XPS to get an idea about its oxidation state [104]. Raman spectroscopy is a powerful tool for studying the structures of polarons and bipolarons, which are associated with the doped conjugated polymers. Electronic absorption spectroscopy (EAS) as well as Raman spectra of doped conjugated polymers such as polythiophene, poly(p-phenylene), and poly(p-phenylenevinylene) with different counter ions have been taken as examples [105]. These analyses have made the correction that polarons are the major species generated by doping in these conjugated polymers in contrast with the previous view that bipolarons are the major species. The change of relative intensity in the Raman spectrum of single-wall carbon nanotubes (SWCNT) is observed for poly(3,4-ethylendioxythiophene)/polystyrenesulfonate/SWCNT nanocomposite sample due to partial doping interaction of SWCNT with the polymer composite [106, 107].

5.4 Electrochemical Property

Electrochemical methods can be used as important evidence for doping in conjugated polymer, since the doping of a conjugated polymer involves redox processes. Details about the electrochemical characterizations of conjugated polymers have already been discussed elsewhere [90, 108]. For studies of electrochemical properties in a three-electrode system, the conjugated polymers are generally prepared by electrochemical polymerization to get a film on the electrode surface.

Chemically synthesized conjugated polymers are generally cast as a thin film on the electrode surface for electrochemical characterization using a simple solution evaporation technique, spin coating, dip coating or pressed pellets. Though various electro-analytical methods can be used to study the electrochemical properties of conjugated polymers, a very popular technique is the cyclic voltammetry and electrochemical impedance spectroscopic technique.

5.4.1 Cyclic Voltammetry

Cyclic voltammetry is favored as an electrochemical characterization technique because the method clearly describes the formation of conjugated polymers as well as indicates the potential range of charging and discharging for the polymers. In principle, depending on the doping conditions, the doped conjugated polymer can store a higher amount of excess charge which can be observed from electrochemical characterization. Characteristic cyclic voltammetry of conjugated polymers has steep anodic waves at the start of charging due to oxidation followed by a broad, flat plateau with an increase of potential. A potential shifted cathodic peak-current, which appears at the negative end of the capacity-like plateau, is generally smaller than that of the anodic peak. During the anodic scan at positive potentials a great amount of charge is consumed by polypyrrole [109], polythiophene or polymethoxydithiophene [110] in the presence of appropriate electrolyte. All these facts can be well explained in the characteristic cyclic voltammetry feature [111] of the electrochemical doping process of poly-3,4-ethylenedioxythiophene using four different electrolytes, Bu_4NClO_4, Bu_4NBF_4, Bu_4NPF_6 and $Bu_4NCF_3SO_3$, in acetonitrile solution (Figure 5.4). Application of an electrochemical approach to poly(3,4-ethylenedioxythiophene) (PEDOT) films on Au electrodes in $LiClO_4/CH_3CN$ solution has explained the mechanism of the p-doping and dedoping process [90]. Moreover, during cyclic voltammetry

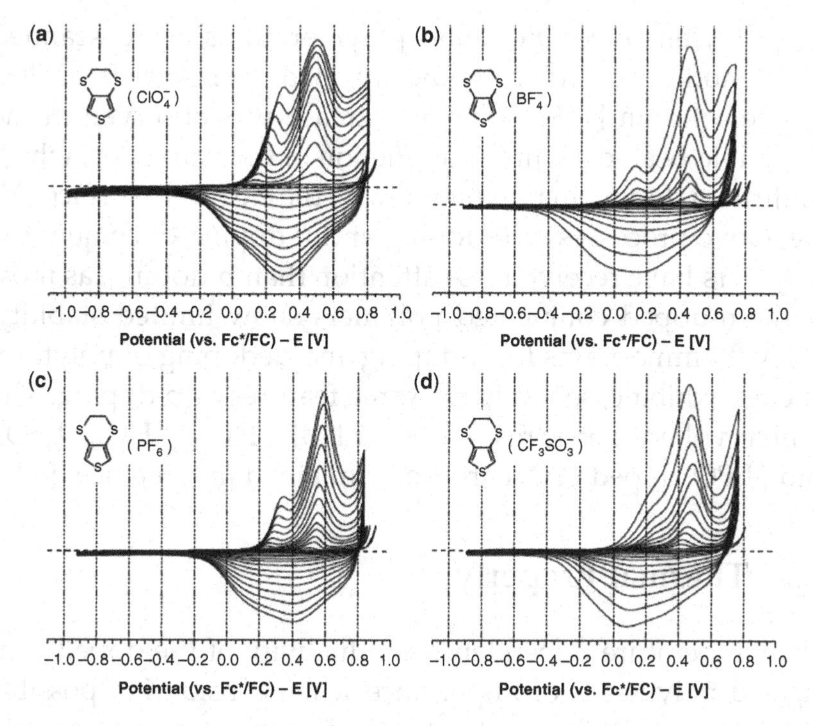

Figure 5.4 Cyclic voltammograms of redox doped poly(3,4-ethylenedithiothiophene) during electrochemical synthesis with (a) Bu_4NClO_4, (b) Bu_4NBF_4, (c) Bu_4NPF_6, (d) $Bu_4NCF_3SO_3$ electrolytes in acetonitrile solvent. Reproduced with permission from ref. [111], Copyright © 2011 Elsevier B.V.

experiments with respect to potential scan rate the doping and dedoping responses are very distinct.

5.4.2 Electrochemical Impedance Spectroscopy

The capacitance of the conducting polymer film measured by electrochemical impedance spectroscopy increases only as the polymer gets oxidized or doped, and the doping process only occurs at certain potentials. As for example, an electrochemical impedance spectroscopy (EIS)-based method is reported to measure the doping level of poly-3-hexythiophene (P3HT), poly-3,4-ethylenedioxythiophene (PEDOT) and polypyrrole (PPy) [112]. In order to represent the ability of n- and p-type doping of various dialkyl substituted

3,4-propylenedioxythiophene polymer films were scanned both anodically and cathodically in a monomer-free electrolyte solution [113]. A linear increase was observed in the reversible peak currents as a function of the scan rates, which confirms a doping in polymer film on the electrode surface. All the above processes of p-doping and n-doping of conjugated polymers have received less attention than p-doping, as most of the n-doped conjugated polymers show limited stability. The voltammograms for n-doping and dedoping of polythiophenes exhibit essentially the same features as p-doping. The similar voltammetric responses for HBF_4, $HClO_4$, HCl, H_2SO_4, and H_3PO_4 doped PANI are well explained in reference [97].

5.5 Thermal Property

The temperature dependence conductivity of the doped conjugated polymer is of importance when considering possible applications within a particular temperature range. Generally speaking, the conductivity of a metal falls as the temperature is raised. This is due to the decrease in electron mobility for band transport with increasing temperature, which enhances the scattering processes by lattice phonons. In simple words, the faster directionless movement of electrons at higher temperature in metal makes them less available to carry the current. In this case, the temperature coefficient of the resistance is positive (PTC). Whereas, if the temperature of an inorganic semiconductor such as silicon is raised the conductivity increases. Such materials are said to have a negative temperature coefficient (NTC). The arbitrary comparison of PTC for silver metal and NTC for iodine-doped polyacetylene is shown in Figure 5.5. The temperature dependence of conductivity for doped conjugated polymer is like that of an inorganic semiconductor but not like metal, even though many conducting polymers show conductivity in the metallic region. Unlike metal the conductivity of all doped conjugated polymers increases with

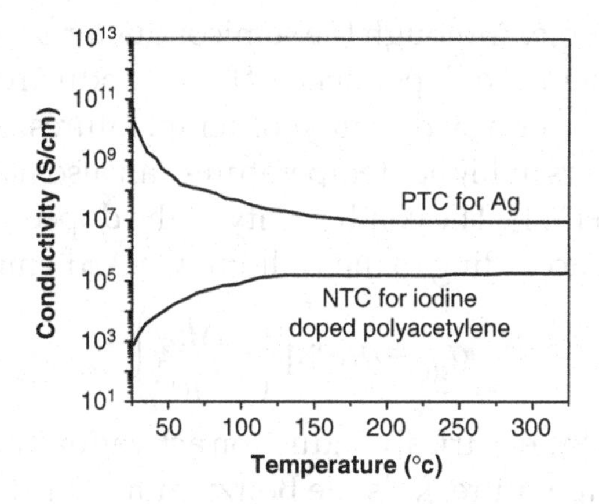

Figure 5.5 The arbitrary comparison of PTC for silver metal and NTC for iodine-doped polyacetylene.

the increase of temperature, which is typical of semiconducting systems. As the conductivity ($\sigma = n\mu e$, where, e-electron n-number of e and μ-mobility of e) in doped conducting polymer depends on the mobility of the electron, here conductivity increases with an increase of temperature. If the upper temperature limit is kept below degradation temperature or critical value, a completely reversible drop in conductivity is observed with decreasing temperature. It can be seen that the conductivity of the halogen-doped polyacetylene decreased as the temperature was decreased for 300–3 K. However, the conductivity generally decreases with temperature for the conductivity of metal like Ag. This is because as the temperature increases the movement of electrons becomes faster and availability of the electrons is decreased to take up or carry the electricity. But in the case of conducting polymer, conductivity depends on the charge (e), electron mobility (μ) and number of responsible electron (n): i.e. $\sigma = n\mu e$. So, with an increase of electron mobility at a higher temperature the conductivity of the doped polymer is also increased. This indicates that the behavior of doping polyacetylene is basically different from that of a metal or classical inorganic semiconductor and that, at high levels of concentration of oxidant, a semiconductor-metal transition is indeed plausible. Although the semiconductor or non-metallic sign in temperature dependence of the conductivity plot can be observed over a wide range of temperatures, some conjugated polymers at higher temperatures can also have the same metallic sign [114]. The conductivity of the doped polymer can be described according to the well-known Arrhenius equation:

$$\sigma_{dc} = \sigma_0 \exp\left(-\frac{\Delta E_\sigma}{k_B T}\right)$$

Where, ΔE is the electrical conduction activation energy, T is the absolute temperature, k_B is the Boltzmann's constant and σ_o is the pre-exponential factor (including the charge carrier mobility and density of states). The fitting of conductivity properties against temperature with the Arrhenius equation reveals the typical semiconductor behavior with negative slope of the $\ln\sigma$ Vs. 1000/T of the doped polymer. The temperature dependence of conducting properties of some common polymers with various dopants, like polyacetylene, polyaniline, polypyrrole, polythiophene and poly(p-phenylene vinylene), should be enough to have an idea about that theory.

5.6 Structural Property

There are many levels of polymer structure like the connectivity of the atoms, three-dimensional shapes due to short-range non-bonded interactions, the shape or conformation of the polymer chains due to long-range non-bonded interactions and degree of arrangement or crystalline structure. Morphology is defined as the study of the form of the doped conjugated polymer and it greatly depends on the structure of the polymer. In the case of conjugated polymers, morphology generally describes the three-dimensional chain conformation and the relationship between chains, as well as the aggregates. Furthermore, morphology includes the physical appearance of polymer particles such as rice grains, spheres, tubules, and fibrils. In the doping process, the dopant counter ions are introduced for stabilizing the charge along the conjugated polymer backbone. The incorporation of such types of dopant counter ions into the rigid delocalized π-electronic structure can create a hindrance. This may cause an undesirable structural distortion, which may affect the crystallinity and morphology of the conjugated polymer.

5.6.1 Crystal Structure

The X-ray diffraction (XRD) structural effects due to doping for various conducting polymers have been well reported earlier [115, 116]. Doping leads to uncertain different crystal forms for the conducting polymers with respect to the undoped crystalline regions of polyacetylene, poly(p-phenylene), poly(p-phenylene vinylene), and polythiophene [117]. Studies on the local crystalline structure of conducting polymers have been helpful in getting detailed information regarding the intrinsic redox state due to doping. A detailed study for polyaniline has been reported [118]. In that study the increase of crystallite size and decrease in interchain separation was observed with increasing protonation or doping due to an increase in three-dimensional delocalization of charge through the polymer chain. As an exception, the intensity of the XRD peaks of the undoped polymer is higher than that of sulfuric acid-doped poly(m-aminophenol), and the peaks totally disappear for the polymer doped with perchloric and phosphoric acids [92]. According to the explanation, this is due to the incorporation of large negative dopant counter ions in the doped poly(m-aminophenol) by breaking the intermolecular hydrogen bonding, which makes the undoped polymer more crystalline.

5.6.2 Morphological Structure

Scanning electron microscopy (SEM) and tunneling electron microscopy (TEM) have initially been employed as tools for the study of morphologies of conducting polymers. However, these techniques provided little information on their electrical properties and chemical structure. Thus SEM and/or TEM with XRD analyses are helpful in getting concrete information about the structural and morphological property of doped conjugated polymer. XRD data with supporting scanning electron microscopic (SEM) and transmission electron microscopic (TEM) analyses have shown that the dopant anion also has a profound influence on the morphology of electrodeposited polypyrrole [119, 120]. The coating of conjugated polymer over the nanomaterials is observed as the monomer is absorbed on the nanomaterials having high specific surface area [121, 122]. The morphological structure is highly dependent on the concentration of the nanomaterials used to prepare the conjugated polymer nanocomposite. As for example, at very low concentration of multi-walled carbon nanotube, polyaniline-coated tubes exist as globular agglomerates while, with an increase in multi-walled carbon nanotube content the morphology is changed from highly aggregated globules to uniformly coated tubules [121]. Similarly, coiled-shape carboxylic acid group functionalized carbon nanotubes are coated with the polyaniline matrix (Figure 5.6) rather than fibered arrangements of the nanotube probably due to the strong doping interaction of carboxylic acid functional groups with polyaniline [99].

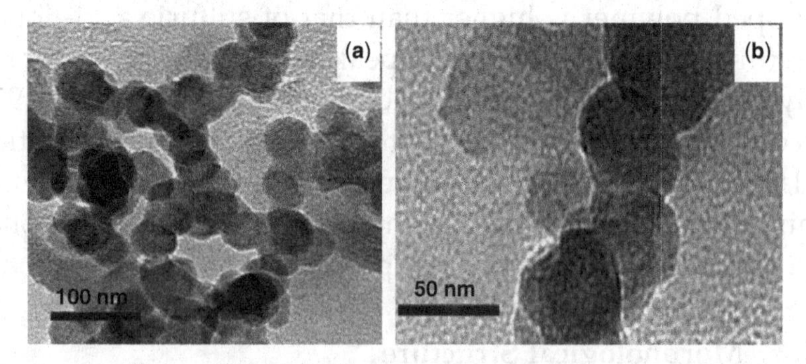

Figure 5.6 SEM images of carboxylic acid functionalized carbon nanotube doped polyaniline nanocomposite. Reproduced with permission from ref. [99], Copyright © 2011 Elsevier B.V.

6

Some Special Classes of Dopants for Conjugated Polymer

6.1 Introduction

The doping property is somewhat specific and universal in its action. That means a particular dopant can only dope a particular conjugated polymer, or it is also able to dope other conjugated polymers through the same type of reactions. In addition, not all dopant is useful for all conjugated polymers, or a single conjugated polymer cannot also be doped by all the available dopant. For a conjugated polymer, doping with specific dopant may have the highest conductivity but it may not have stability within the polymer matrix. So, the property of the doped conjugated polymer depends on the doping interaction between the particular dopant and that polymer. Various types of dopants are used successfully to dope different types of π-conjugated polymers. Among those, some dopants like iodine, halogens, halides, protonic acids, etc., are very important as well as special for some conjugated polymers like polyacetylene, polyaniline, polypyrrole, polythiophene, etc. In the case of such types of dopants the interaction with the different types of conjugated polymers is more or less similar. As for example, the protonic acid always acts as a neutral p-type dopant and it protonates the basic heteroatom center of the conjugated polymer chain. Therefore, some special types of dopants and their doping interactions with the different types of conjugated polymers are generalized in this chapter.

6.2 Iodine and Other Halogens

6.2.1 Principle

Halogen doping in polyacetylene has been studied with great historical interest because of the metallic properties of the doped polymer [33]. The doping was first introduced in 1977 with the findings of Shirakawa *et al.* [12] as iodine vapor doping on trans-polyacetylene. Polyacetylene also shows a dramatic increase in electrical conductivity when the polymer is doped with controlled amounts of other halogens viz., chlorine and bromine. However, there is almost no example for the use of fluorine as dopant in conjugated polymer probably due to the toxicity of the reagent. Here halogens are acting as p-type, inorganic, redox dopant. It is now well established that the exposure of polyacetylene to a halogen dopant leads to the formation of a positively charged polymeric complex and a negatively charged halogen counter ion. Halogens oxidize the conjugated polymer by accepting the π-electron from the C center (polyacetylene, poly-p-phenylene, poly-p-phenylene vinylene, etc.) or more electronegative atom N, S, etc., if present (polyaniline, polypyrrole, polythiophene, etc., and their derivatives). The studies based on Raman spectroscopy have clearly shown that the polyacetylene chain acts as a positively charged polycation counterbalanced by I_3^- or little I_5^- ions but not I^- ions. However, for other halogens there is no such evidence, i.e., the counter anion is Cl^- or Br^-.

Due to halogen doping polarons can generally be found in conjugated polymers but most of them do not carry solitons. The positively charged polymeric radical cation (also called a "polaron") normally has a very low mobility because of Coulombic attraction to its counter ion (I_3^-), and a local change in the equilibrium geometry of the radical cation relative to the neutral molecule. Charge transport in such types of polaron-doped polymers occurs via electron transfer between localized states being formed by charge injection on the chain. Here some localized soliton also results due to jumping of an electron in localized states of an adjacent polymer chain.

6.2.2 Doping Technique

Chlorine and bromine are available as gas and iodine is available as sublimated solid at room temperature. The doping of halogens to the conjugated polymer is possible by gas- or liquid-phase chemical doping or the electrochemical doping process. In the chemical gas-phase doping process, the conjugated polymer is exposed to halogen within a closed chamber at controlled temperature and pressure. The iodine doping set-up is as simple as shown in Figure 6.1. In a typical doping

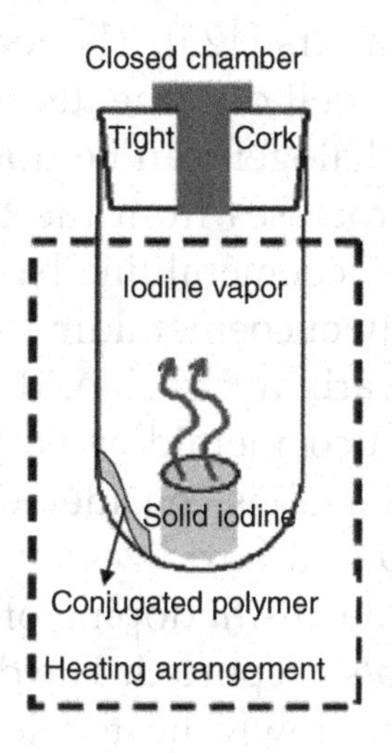

Figure 6.1 The simple iodine doping set-up.

method, iodine crystals are placed in a glass beaker and the polyaniline films to be doped are kept above this glass beaker. As the beaker is slowly heated, the iodine vapors are absorbed into the polymer films and thus the color of the polyaniline thin film changes from light to dark brown due to doping [62]. For gaseous bromine and chlorine the pressure is monitored directly at room temperature using gaseous doping set-up, whereas for sublimated solid generally the iodine temperature is maintained at atmospheric pressure. As for example, the halogen doping effect was demonstrated by exposing the polyacetylene film in halogen vapor at known pressure and temperature [12]. The doping of polyacetylene film was reported by using both gas-phase chemical doping from iodine vapor and solution-phase chemical doping from 5-mM solution of iodine in hexane [123]. The electrochemical doping of polyacetylene film on anode was employed in aqueous KI or KI/I_2 solution [124]. Like the polyacetylene-iodine system, the donor-acceptor interactions between π-conjugated polymers (e.g., polythiophene, polypyrrole, polyphenylene, and their derivatives) and various halogens can be considered as a useful method for doping. For example, the halogen doping of poly-p-phenylene was done at 30 to 65°C by exposing it in various halogen vapors [125]. The iodine doping in polyaniline has also been well documented in the literature. The oxidative doping by halogen can be done through chemical or electrochemical processes from the totally reduced form of polyaniline, polyleucoemeraldine base. The chemical oxidative doping of polyleucoemeraldine was performed with chlorine or iodine in carbon tetrachloride solution. The oxidative doping of polyleucoemeraldine with Cl_2 is illustrated in Scheme 6.1 [32]. When Cl_2 is considered both as the oxidant and the dopant, H_2O_2 only oxidizes polyaniline doped with the acidic solution. The *in-situ* doping of conjugated polymer by halogen vapor is also reported. The doping of polyaniline was demonstrated in slowly heated iodine vapors as well

Polyleucoemeraldine base

Polyemeraldine salt

Scheme 6.1 The oxidative doping of polyleucoemeraldine with Cl_2. Reproduced with permission from ref. [32], Copyright © 2003 Elsevier B.V.

as *in situ* by simultaneous spraying of the monomer and the dopant iodine vapor during plasma polymerization [126].

6.2.3 Property

The iodine doping in polyacetylene is more popular than the other halogens due to better results. The room temperature conductivity of pure undoped polyacetylene films varies from 10^{-5} S/cm for the trans form to 10^{-9} S/cm for the cis form, and high conductivity >20,000 S/cm has been reported for partially orientated iodine-doped polyacetylene [127]. The doping reactions result in the formation of polymer complex salt which may be generally indicated as $(CHX_y)_x$, where y represents the moles of dopant per polymer repeat unit. The changes in conductivity of polyacetylene upon exposure to iodine, as a function of the time of reaction and the concentration of reactants, are shown in Figure 6.2 [128]. The conductivity increases when the concentration of the iodine increases with an initial sharp jump in conductivity. The remarkable and unique temperature dependence conducting behavior of bromine and iodine-doped polyacetylene is shown in Figure 6.3 [58]. It can be seen that, unlike metal conductor, the conductivity of the halogen-doped polyacetylene decreased as the temperature is decreased for 300–3 K.

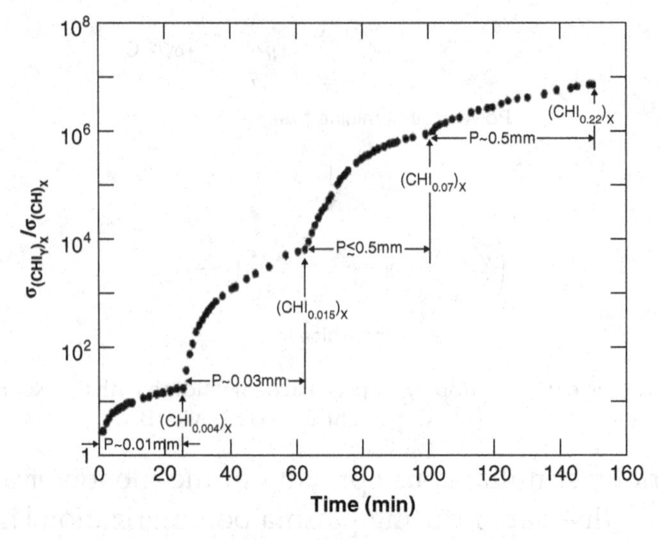

Figure 6.2 Effect of conductivity on doping time when the polyacetylene film is exposed in 1 torr constant pressure of iodine vapor. Reproduced with permission from ref. [128], Copyright © 1978 American Institute of Physics.

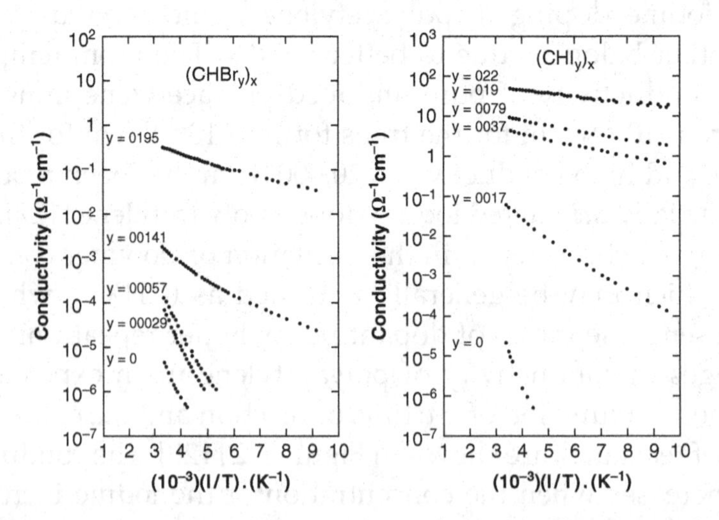

Figure 6.3 Temperature dependence conducting behavior of bromine- and iodine-doped polyacetylene. Reproduced with permission from ref. [128], Copyright © 1978 American Institute of Physics.

In general, upon doping the cis-polyacetylene isomer shows higher conductivity (about 2–5 times) than that of the trans isomer, although room temperature conductivity of undoped trans isomer is greater due to better planarity [129].

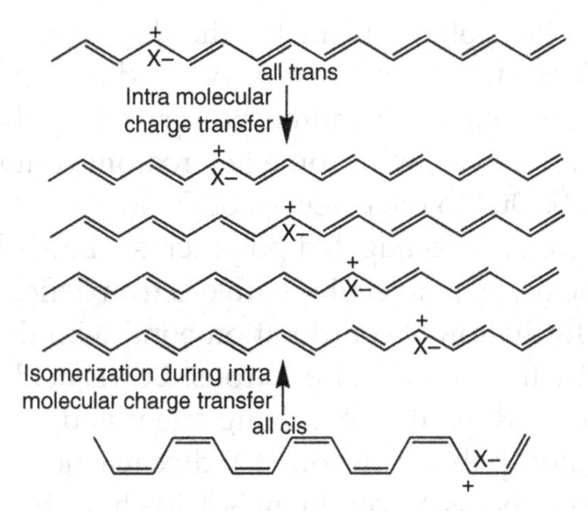

Scheme 6.2 Isomerization of doped (all cis) polyacetylene during doping.

Experimentation suggests that isomerization occurs during doping and both isomers give the same product after doping as shown in Scheme 6.2.

The iodine-doped cis isomer is higher conducting probably due to this type of bond rotation, which makes interchain electron transfer easier.

The halogen-doped conjugated polymers only have an historical importance. They were first invented for the polyacetylene system, and since then have been tried for almost all the conjugated polymers. The halogen gases are not very stable inside the conjugated polymer matrix and the doped polymer loses its conductivity just by environmental exposure alone. Thus, halogen doping of conjugated polymers nowadays has become of less interest to researchers.

6.3 Halide Doping

6.3.1 Principle

Doping with halides, which is Lewis acid in nature, is another way to drive charge transfer within conjugated polymer for electrical conduction. Metal and nonmetal halides are very good inorganic dopant electrolyte and can easily produce the ions in suitable polar solvent for the doping of conjugated polymer. These Lewis acids are very good dopant for conjugated polymers, e.g., polyaniline, polypyrrole, polythiophene and their derivatives, with some heteroatom-containing loan pair like N, S. Due to the presence of the lone electron pair on the heteroatom the conjugated polymer is a Lewis base. Thus the Lewis acids, i.e., molecules with electron deficit center, are capable of formation of coordination bond with those heteroatoms via their lone pair of electrons. Lewis acid halide can act as a p-type dopant or oxidizing agent and it can oxidize the conjugated polymer through redox reaction assisted by an acid-base process. Several Lewis acids have been tested as polyaniline dopants, for example, BF_3, $AlCl_3$, $SnCl_4$ and $FeCl_3$. The reduced form of the dopant acts as a counter anion to maintain the electroneutrality inside the conjugated polymer matrix. As for example, doping of the polythiophene or polypyrrole systems occurs by oxidation of $FeCl_3$, and here the polymer becomes positively charged associated with $FeCl_4^-$ counter anion [130]. The doping occurs for the conjugated polymers without having heteroatom in the polymer chain through simple oxidation reaction. The scheme for such doping in trans-polyacetylene by gaseous AsF_5, which is believed to dope the polymer to form the hexafluoroarsenate ion and arsenic trifluoride can be shown as [129];

$$3\,AsF_5 + 2(-CH{=}CH-)_x \rightarrow 2(-CH-CH^+-)_x + 2\,AsF_6^- + AsF_3$$

AsF_3 shows almost a similar transformation process. However, the doping mechanism may also be a non-redox one for the halides with very small sized metal ion. In such a case as in protononic acid doping, the small cation adds withis added to the heteroatom of the conjugated polymer and makes the polymer positively charged with the cation of the halide dopant. There is a negative part of the dopant, i.e., halide anion remains with the positively charged center in conjugated polymer chain as counter ion. Such a doping type is observed for LiCl dopant in polyaniline co-doped with HCl [18].

6.3.2 Doping Technique

Halide doping can be performed chemically in the vapor phase or solution phase. In vapor phase chemical doping, a polymer like polyacetylene is exposed to a gaseous halide, e.g., BF_3, AsF_5 or SbF_5. Investigators have established that counter ions for these acceptor-doped polyacetylene include BF_4^-, SbF_6^- and AsF_6^-. For the chemical solution phase doping of $FeCl_3$ in polypyrrole or polythiophene (and their derivatives) 1 M $FeCl_3.6H_2O$ can be used. In the doping process the counter ion which is used to form the polymer charge transfer salt is $FeCl_4^-$. These electrochemically synthesized undoped polypyrrole films were then doped with metal halide dopants viz., $PdCl_2$, $CuCl_2$, $NiCl_2$, $MnCl_2$, $CoCl_2$, $FeCl_3$, and $ZrCl_4$, by placing them in the respective salt solutions (slightly acidic with 2 mM HCl in the case of $PdCl_2$) [131]. Emeraldine base can be obtained by heating the emeraldine salt with ammonia solution. In another method, separate salt solution was prepared by dissolving the MX conducting (M=Metal and X=Halide) in distilled water, and the solution was slowly added to the precooled polymer solution with constant stirring. Then the mixture solution was dried in an oven at 60–70°C temperature to get the polymer in the powder form. By this method doping of the polyaniline was done with KBr and after doping a thin film of polyaniline was also prepared [132]. The *in-situ* doping of halides is also common. As for example, *in-situ* halide-doped polypyrrole was chemically synthesized in aqueous medium containing a mixture of metal halide (Co, Ni) with or without HCl using ammonium persulfate oxidant [133]. *In-situ* electrochemical doping is also possible for the various halides as they are good electrolyte. For example, dopants like $C1O_4^-$, $FeC1_3$ and $IrC1_6^{2-}$ were also investigated for acceptor doping in the polyacetylene system [127, 134].

6.3.3 Property

The difference between AsF_5 and iodine may be due to a greater solubility of iodine in polyacetylene and a lower fully doped level, such that the contrast in permeability between the doped and undoped regions is much less than with AsF_5 [129]. The attractive redox electrochemical behavior of halide-doped conducting polymers has drawn the attention of researchers for the use of these polymers as conducting, electroactive materials in various applications like electromagnetic interference (EMI) shielding, rechargeable batteries, chemical sensors, and corrosion protection. In view of the above potential applications a number of metal halide salts like LiCl, $MgCl_2$, NaCl, KCl, $LiClO_4$, $LiBF_4$, $Zn(ClO_4)_2$, etc., have been used as dopant for polyaniline and their derivatives [135–137].

6.4 Protonic Acid Doping

6.4.1 Principle

The spectroscopic properties of metal halide doped polyaniline are significantly different from those of protonic acid doped polymer due to the different types of chemical interactions in both systems. The protonic acids are the Brönsted acids, which can easily give the H^+ ion in the aqueous phase. On the other hand, the conjugated polymer with heteroatoms in the chain can behave as weak Brönsted base, which can take up proton in aqueous medium. The doping of conjugated polymers by protonic acids is an acid-base reaction in the Brönsted sense and consists of the protonation of the base form of the conjugated polymer containing heteroatom with a sufficiently strong protonic acid (Scheme 6.3). This is a p-type neutral doping, which does not donate or take up electron from the π-conjugated polymer. Generally these protonic acids protonate the polymer containing basic functional

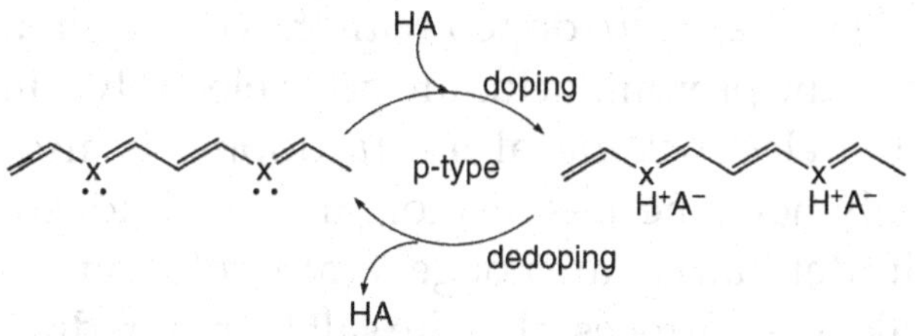

Scheme 6.3 Protonic acid doping mechanism.

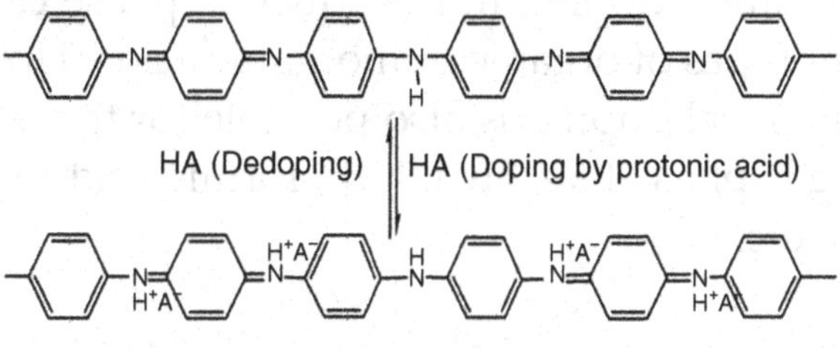

Scheme 6.4 Protonic acid doping in polyaniline.

group (which can donate electron pair) and salt formation occurs due to having the conjugate base as counter ion inside the polymer matrix. Protonic acid doping can be termed as neutral doping, as no electron transfer occurs from or to the conjugated polymer. Such neutral dopants show more importance in the case of polyaniline and its derivatives over redox dopants due to the easy doping procedure, good reproducibility and good reversibility as well as better conductivity (Scheme 6.4). The processability of polyaniline and its derivatives can be improved by doping with functionalized protonic acids [138] provided the protonic acids are sufficiently strong for polymer protonation.

6.4.2 Doping Technique

Various organic as well as inorganic protonic acids are used as neutral dopant to dope polyaniline and its derivatives. For polyaniline and its derivatives, the polymers are generally synthesized in aqueous acid medium, and so the resultant polymers became *in-situ* doped with the corresponding acids. Similarly, as the protonic acids are good electrolyte in aqueous solution the electrochemical *in-situ* doping is also common for the polymer. The *in-situ* inorganic acid-doped polyaniline and its derivatives are not generally solution processable, and to make them processable the salt form is reduced to base form by dipping in ammonia, phenyl hydrazine or hydrazine solution. After processing the dedoped polyaniline the chemical solution doping in the aqueous phase can be done by different types of organic or inorganic acids. However, the gaseous chemical doping is also possible for the acids which are in the gas phase like, HCl, HBr, formic acid vapor, acetic acid vapor, etc.

6.4.3 Property

Processability is the main problem for *in-situ* inorganic acid-doped polyaniline. A general observation is that the better dopant after the dedoping and processing of *in-situ* inorganic acid-doped polyaniline is the protonic acid which is used during the synthesis of the polymer. This fact can be explained as the "memory effect" [139, 140] obtained during the polymerization from the acid medium. The reason why the doping ability of acid used in polymerization or "memory effect" is the best is that during synthesis the polymer is attached with the counter ion of the dopant so that the space is best fit or available for the counter ion after processing. To improve the solubility of conducting polyaniline in the doped form various organic protonic acid dopants are used with very big counter ion. Such examples are shown in Table 6.1 [139, 141]. The use of this big dopant increases the processability of the doped polyaniline but it decreases the conductivity of the polymer. Another problem is that film casting is not possible from the solution of the doped form. For this reason most of the researchers reported the conductivity of polyaniline doped with those organic acids in pellet form. Small chain protonic organic acids are also used for the doping of polyaniline. Here the conductivity of the doped polyaniline is less than that of the inorganic acid-doped polyaniline. In the case of polyaniline derivatives all the above acids and oxidants have been tried for doping polymer by various research groups. The conductivity of polyaniline and its common derivatives doped with neutral acid is shown in Table 6.2. From their experiments it can be concluded that HCl is the best dopant for polyaniline and its derivatives in terms of conductivity.

Table 6.1 Solubility and conductivity of emeraldine salt with (SO_3^-R) counter ion (reproduced with permission from reference 141 Copyright © 1992 Elsevier Inc.)

R	σ (S/cm)		Solubility[b]				
	Pellet	Film[a]	Xylene	CH_3Cl	m-Cresol	Formic acid	DMSO
C_6H_{13}	10		O	O			
C_8F_{17}	19		O	O			
$C_8F_{17}COOH$	2.7		--[c]				
C_8F_{17}	3.7		--[c]				
(L,D)-Camphor	1.8			Å	Å	Å	O
4-Dodecylbenzene	26.4		Å	Å	Å	O	
o-Anisidine-5-	7.7×10^{-3}	100–400			O	O	O
p-Chlorobenzene	7.3	100–250			O	O	O
4-Nitrotoluene-2-	5.7×10^{-2}				O	O	O
Dinonylnaphthalene	1.8×10^{-3}		O	Å	Å		
Cresol red	2.2×10^{-4} [d]				O		
Pyrrogallol red	1.2×10^{-1} [d]				O		
Pyrrocatechol violet	1.9×10^{-1} [d]				O		

[a]Films were cast from concentrated solution; [b]O = soluble at room temperature 'Å = very soluble at room temperature; [c]Soluble in perfluoroalkanes, e.g., perfluorodecaline; [d]Pressed at 165 °C

Table 6.2 Conductivity of polyaniline and its common derivatives.

Sl. No.	Polymer	Dopant	Conductivity (S cm^{-1})
1	Polyaniline	Acid	50
2	Poly (*o*-methyl aniline)	Acid	0.30
3	Poly (*o*-ethyl aniline)	Acid	0.68
4	Poly (*o*-methoxy aniline)	Acid	0.14
5	Poly (*o*-bromo aniline)	Acid	10^{-8}
6	Poly (*o*-iodo aniline)	Acid	10^{-8}
7	Poly (*o*-chloro aniline)	Acid	10^{-6}
8	Poly (*o*-amino aniline)	Acid	10^{-9}
9	Poly (*o*-hydroxy aniline)	Acid	10^{-10}
10	Poly (*m*-hydroxy aniline)	Acid	10^{-8}

6.5 Covalent Doping

During doping covalent bond formation may occur between the dopant and conjugated polymer or the inside of the conjugated polymer. This type of covalent bond formation during the doping process, which is termed covalent doping, influences the conductivity of the doped conjugated polymer. In some cases this chemical modification provides a more extended conjugated structure which increases conductivity of the doped polymer. In other cases this reaction interrupts conjugation and, correspondingly, limits the conductivity of the doped conjugated polymer complexes [142]. The doping of poly(p-phenylene) with SO_3 from oleum at high temperature is an example. The chemical modification (B in Scheme 6.5) results in five-membered chains due to the covalent addition of a SO_3 with the neighboring aromatic chain. Such chemical modification leads to a loss of conductivity compared to that of simple sulfuric acid-doped polyphenelyene

(A in Scheme 6.5), probably due to disturbance in planarity of the conjugated system, and hence electron delocalization throughout the conjugated system. These side reactions suppress the resultant conductivity. Another system of interest is poly(p-phenylene sulfide) doped with AsF_5 [143–146]. Here, two types of covalent doping is possible, one intrachain bridging to form thiophene rings (B in Scheme 6.6) and the other an interchain bridging to form crosslinking polymer network (C in Scheme 6.6). The bridged and crosslinked polymers are formed charge-transfer complexes with the dopant as well since there is evidence that these altered polymers are highly conducting. The self-doped conjugated polymer (discussed earlier) in that sense can also be termed as covalent doped polymer since the covalent bonding is present in between the dopant and the conjugated polymer.

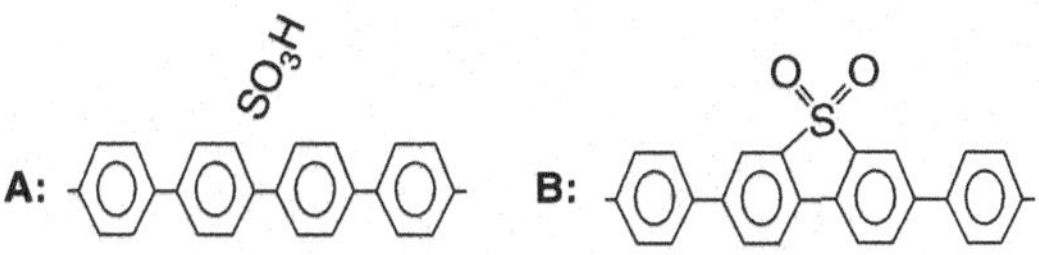

Scheme 6.5 Structure of chemical modified poly(p-phenylene) during doping with SO_3 from oleum at high temperature [142].

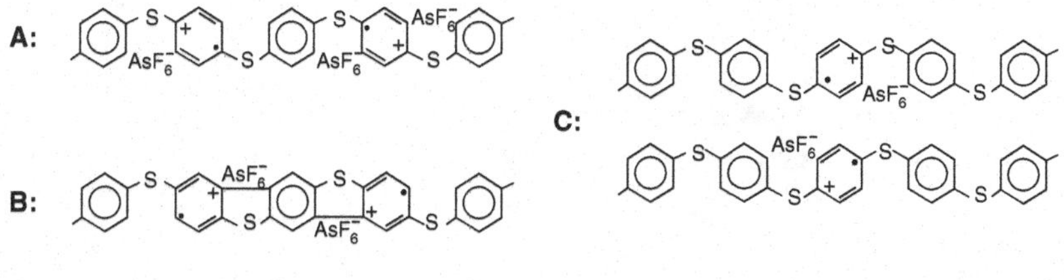

Scheme 6.6 Covalently doping in poly(p-phenylene sulfide) during doping with AsF_5 [142].

7

Influence of Dopant on the Applications of Conjugated Polymer

7.1 Introduction

Conjugated polymers with electrochemical, electrical and optical activity have emerged as the electroactive materials to replace inorganic semiconductors in the electronic and electrical industry, as well as offering themselves as important materials for the optoelectronic industry. The possible applications for conducting polymers in undoped as well as in doped states are of recent research and development interest. However, conjugated polymers are highly susceptible to chemical or electrochemical oxidation or reduction, i.e., doping, which alters the optical and electrical properties of the polymer. It is possible to precisely control the above properties by controlling the dopant type and doping level. There are two main groups of applications for doped conjugated polymers. The first group utilizes their conductivity as its main property and the second group utilizes the electrochemical activity of the conjugated polymer. Whatever

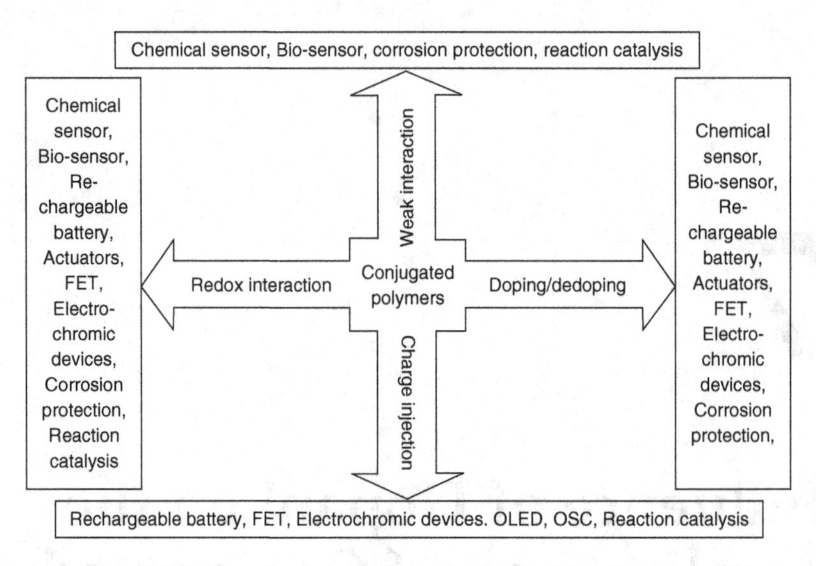

Figure 7.1 Doping mechanism of the conjugated polymer related to application field.

their field of application, doping has an important role in the conductivity or the electrochemical activity of the conjugated polymer. As described in Figure 7.1 [79], the doping mechanism of the conjugated polymer is also closely related with the application fields. In this section the influence of dopants on the various applications of conjugated polymer is briefly discussed.

7.2 Sensors

7.2.1 Chemical Sensors

Chemical sensors are probably the main application for conjugated polymer. It has been established that an interaction with an analyte which causes the doping for conjugated polymers influences the optical properties of the conjugated polymer, and that characteristic can be successfully used for optical sensor devices. As for example, a polyaniline-based optical pH sensor, ozone, NO_2 and H_2S gas sensors are employed on the basis of various colors in different redox states of polyaniline depending on the doping interaction with the analyte.

The change of optical and electrical properties of conjugated polyaniline occurs due to dedoping interaction of the emeraldine salt with NH_3 gas [133] which decreases the counter ion mobility and polaron density. Therefore, the color is altered from green to blue. The right conjugated conducting polymers which are extremely sensitive to structural perturbations in the electronic network is a cause for self-amplifying their fluorescence quenching response by the absorbtion of external analyte. This structural perturbation of conjugated polymer leads to a strong modification of its electronic band structure, and those polymers can be used as good fluorescent sensor material [147]. An interaction of analyte of oxidized/reduce or protonated/deprotonated conjugated polymer can also be measured by UV-vis or photoluminescence spectroscopy. The change of electronic spectrum of polyaniline is explained by the different degree of protonation of the imine nitrogen atoms in the polymer chain as a result of the change in pH [148].

The electrical conductivity of doped conjugated polymer depends on the concentration of incorporated dopant ions. If their doping level alteration can be possible by transferring electrons from or to the analytes, it can cause changes in resistance as well as work function of the sensing material. Therefore in such cases sensing by conducting polymer is possible, as conductivity of the conjugated polymer is changed reversibly and rapidly at ambient temperature by adsorption and desorption of volatile chemicals. It is observed that nucleophilic gases (H_2S, NH_3, N_2H_4, H_2, etc.) cause a decrease in conductivity for most widely used conducting polymers in gas sensing applications like polythiophene, polypyrroles, polyaniline and their derivatives or composites due to n-type doping or dedoping. However, the mechanism of ammonia by the HCl-doped polyaniline is somewhat different. Here, the analyte ammonia gets attached with the dopant H^+, and due to the weak doping effect of the resultant NH_4^+ ion the conductivity of the film is decreased. A decrease in conductivity due to dedoping of conducting polymers like polypyrrole, poly(3-hexylthiphene) and polyaniline by hydrazine vapor was also reported. On the other hand, electrophilic gases with higher electron affinity than the conducting polymer (NO_x, PCl_3, SO_2, O_2, etc.) show the p-type doping effect by increasing the number of charge carriers in conducting polymer through oxidative doping. For this reason the conductivity of the conjugated polymer may increase or decrease depending on the nature of the analyte. SO_2 or NO_2 gas was found to decrease the resistance in polyaniline [149] and poly-3-hexylthiophene [143, 144]. On the other hand, due to an oxidation by NO_2 an increase in resistance was observed for emeraldine salt nanofiber [145]. The electron donation or withdrawal by the analyte vapors having Lewis acid/base characteristics leads to conductivity changes in the sensor film of doped conjugated polymer ("primary doping") due to some so-called "secondary doping" interaction with the analyte. These interactions, which are weak physical interactions with conducting polymer for the non-reactive volatile organic compounds such as chloroform, acetone, aliphatic alcohols, benzene, toluene, and some other volatile organic compounds (VOCs), not only change the oxidation levels of conjugated polymers, but also influence the conductivity. Recently, the sensing of aliphatic alcohol vapor by sulfuric acid doped poly(m-aminophenol) is reported [150]. Here, the –OH groups of the alcohol molecules get hydrogen bonded with the free phenolic –OH groups of the polymer molecules. So, the conductivity of the doped polymer is further increased due to an increase in the weak electron flow throughout the polymer chain [151], which can be termed as secondary doping or, preferably, an induced doping effect. The same mechanism is explained [152] for NH_3 vapor sensing by HCl-doped poly(m-aminphenol)-silver nanocomposite, as shown in Scheme 7.1. A similar effect can also be observed for CO gas by doped polyaniline film. According to the explanation of Bai and Shi [153], the increase in conductivity is observed due to a net increase of positive charge carriers in the polymer backbone by the interaction

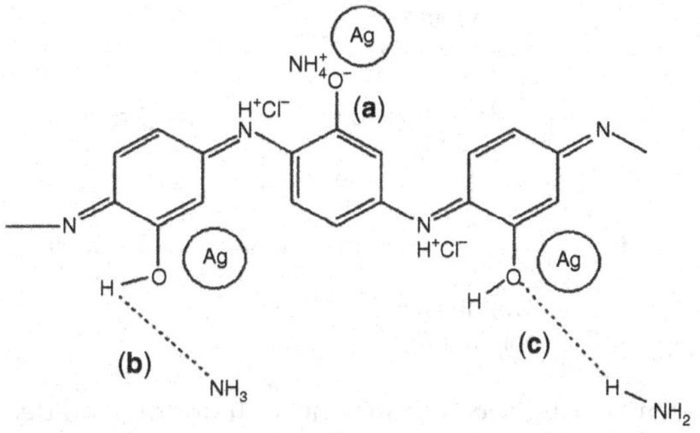

Scheme 7.1 Possible interactions of ammonia vapor with the hydrochloric acid-doped poly(m-aminphenol)–silver nanocomposite. Reproduced with permission from ref. [152], Copyright © Springer Science+Business Media, LLC 2010.

of stable resonating structure of CO as $^{+}C=O^{-}$. In the case of chlorinated hydrocarbon sensing by acid-doped polyaniline similar phenomena were observed [154].

7.2.2 Biosensors

Compatability of conducting polymers with biological molecules in neutral aqueous solutions can be used for biosensing applications with good detectability and fast response. As shown in Figure 7.2 [9], the conjugated polymer can be reversibly doped and dedoped electrochemically in the presence of bioelement. The interaction can cause significant changes in conductivity of the film that can be used as a signal for the biosensing. In the conducting polymer-based amperometric biosensors the redox reaction catalyzed by an appropriate enzyme takes place due to the electron transfer through the polymer layer. For this reason the change of electronic properties of the conjugated polymer can be detected. The sensing of GOD on polypyrrole using redox enzymes has reported a fivefold increase in current response. A similar effect is observed for the sensing of GOD by polyaniline films [155]. Potentiometric biosensors with conjugated polymer can be produced using pH sensitivity of polymers, as the neutral doping and dedoping process of the conjugated polymer depends on the pH of the medium. So, by

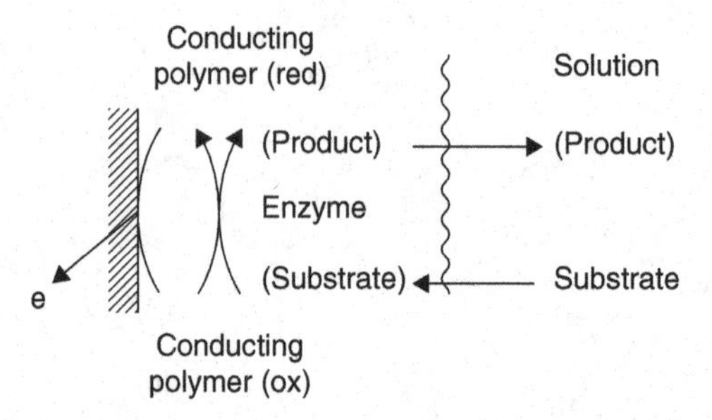

Figure 7.2 Electron transfer mechanism (through doping and dedoping) in conjugated polymer-based biosensor. Reproduced with permission from ref. [9], Copyright © 2002 Elsevier Science B.V.

changing the pH of the environment the redox potential of the conjugated polymer can be altered. Such a type of principle is used for NH_3 biosensor using a polypyrrole sensitive layer [156, 157]. This interaction of NH_3 with polypyrrole was utilized for a design of a potentiometric biosensor of urea with urease immobilized on an electrodeposited polypyrrole layer. A change in redox potential and/or pH of the biological environment on the conjugated polymer matrix can change over several orders of magnitude the response of electronic conductivity of conjugated polymers. Biosensors based on the conductivity response of conjugated polymers have been constructed for penicillin, glucose, urea, lipids and hemoglobin [158, 159]. The Langmuir-Blodgett films of polyaniline and polypyrrole have been used to fabricate a glucose biosensor [160, 161].

7.3 Actuators

The volume change of conjugated polymer by electrical stimulation has also been exploited in electrochemical–mechanical type devices like drug delivery devices and artificial muscle. The swelling of counter ion by the conjugated polymer increases its volume up to 30% [162]. For example, the actuating property of polypyrrole that releases drugs from reservoirs covered by thin bilayer has been reported by applying a small potential to the polypyrrole [163]. Artificial muscle is constructed using two layers of conjugated polymer separated by a nonconductive material. When current is applied across the two conjugated polymer films, one of the doped conjugated layers releases the dopant counter ion and the other conjugated polymer layer accepts the dopant ions. The inflow of dopant counter ions creates an expansion for conjugated polymer which accepts it, whereas the volume reduces for the conjugated polymer film which donates it [164, 165]. The combined effect is translated into a mechanical force that bends the polymer towards the conjugated polymer which is reduced in volume, and it has been compared to the mechanisms in natural muscles (Figure 7.3). The actuation effect

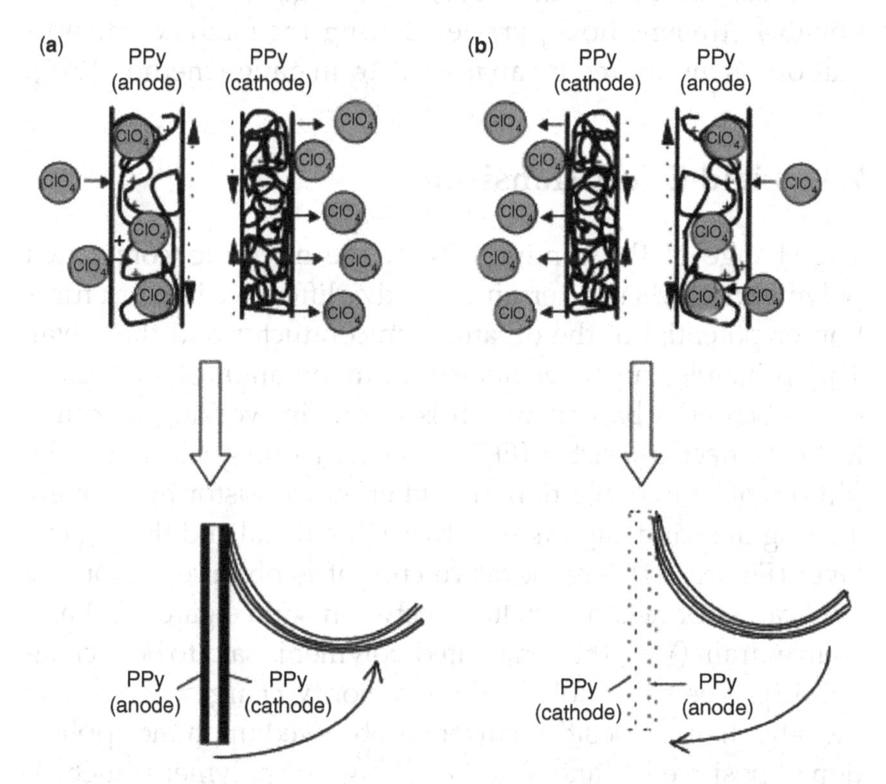

Figure 7.3 Working principle of artificial muscle devices through reversible doping and dedoping of polypyrrole-ClO_4 system. Reproduced with permission from ref. [165], Copyright © 2003 Elsevier Science B.V.

of the polymer depends on the concentration of $LiClO_4$ for a bilayer actuator device based on polypyrrole. These artificial muscles have been exploited for use as actuators for important biomedical applications like steerable catheters for minimally invasive surgery, micro-pumps and valves for labs-on-a-chip, blood vessel connectors, micro-valves for urinary incontinence, etc. Polypyrrole and polyaniline and their composites have been used to fabricate such devices. In other studies polyaniline–carbon nanotube and polyaniline–carbon nanotube–polypyrrole nanocomposites have been employed as actuators. Due to the addition of carbon nanotubes to the polyaniline or polypyrrole fibers an increase of electromechanical actuation was observed [166–169]. The mechanical properties of conducting polymers, e.g., polypyrrole and poly(3,4-ethylenedioxypyrrole), during their redox transformations is monitored by an *in-situ* strain gauge method [170].

7.4 Field Effect Transistor

The change of the conductivity at the electrode/conjugated polymer contacts is determined by the difference in work function or potential of the organic semiconductor and the metal. This principle can be attributed to modulation of the height of the Schottky barrier, which is the main working principle in field effect transistor (FET) devices. Doping can affect the physics of charge injection in field effect transistor by strongly altering the band alignment between the metal and the organic layer (Figure 7.4). The negative current is observed upon the application of negative voltage between source-gate (V_{SG}) and source-drain (V_{SD}), the conjugated polymer is said to be p-channel or p-type since holes are the majority charge carriers. On the other hand, a positive current is observed upon the application of positive V_{SG} and V_{SD}, the conjugated polymer is n-channel or n-type since the electrons are mobile [171]. The organic FETs typically show p-type, but not n-type, conduction as these materials are strong for the trapping of electrons but not holes, except for a few special high-electron-affinity or low-band-gap organic semiconductors. The charge injection in poly(3-hexylthiophene) field effect transistors with Pt and Au electrodes has been reported as a function of annealing in a vacuum [172]. The FET-based devices, which are interesting for chemical sensor design, are polymeric field effect transistors (PolyFET). When the conjugated polymer interacts with gaseous species it can act either as an electron donor or an electron acceptor depending upon the oxidizing and reducing character of the gas on that particular conducting polymer. This interaction of an analyte with

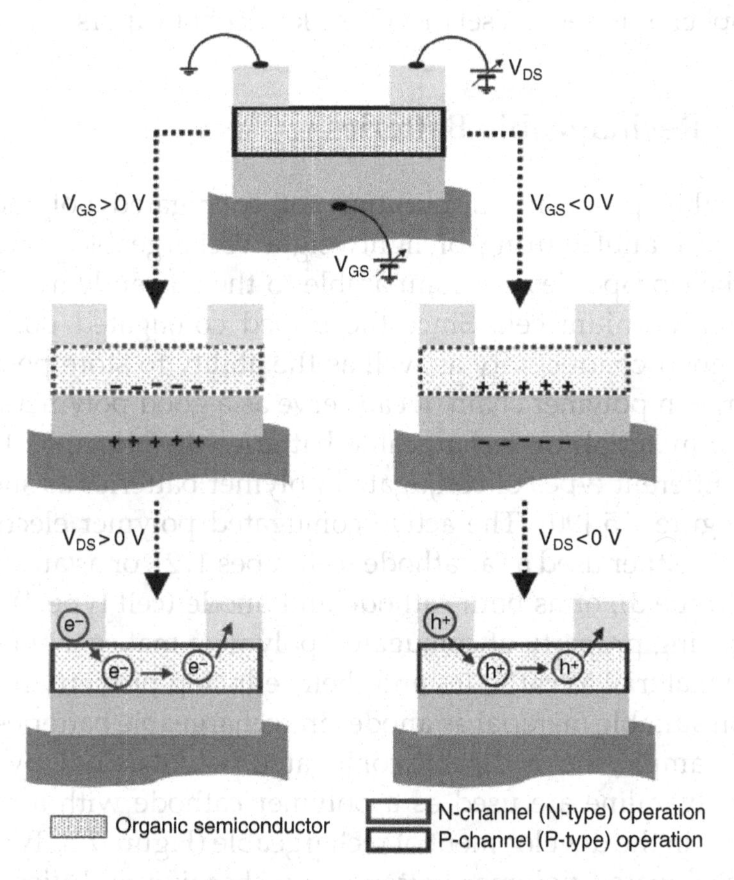

Figure 7.4 Schematic of p- and n-type FET operation. Reproduced with permission from ref. [171], Copyright © 2007 Elsevier Science B.V.

the conjugated polymer layer of PolyFET affects the conductive channel where sensing events occur at the gate or gate/insulator boundary and indirectly modulate the current by capacitive coupling. This is because the charge carriers in conjugated polymers is a measure of mobility of "holes" and "electrons" respectively, and their injection by the analyte may cause the change in the overall voltage of the device. This is a consequence of the density of states of the conjugated systems which dynamically change upon introduction of charge through doping. The conjugated polymer-based gas sensing FET devices are well known as CHEMFET devices. Torsi *et al.* [173–176] have developed gas sensitive CHEMFET using alkyl- and alkoxy-substituted polythiophenes to sense a set of VOCs like alcohol vapors.

7.5 Rechargeable Batteries

A highly promising application for conjugated polymer is in the manufacturing of lightweight rechargeable batteries as their properties are comparable to the currently available nickel-cadmium cell. Since the doped conjugated polymer has good conductivity as well as the ability to store positive charge in polymer chain, it can serve as a good polymer electrode material for rechargeable batteries. In principle, there are different types of conjugated polymer batteries as shown in Figure 7.5 [90]. The active conjugated polymer electrode can be either used as a cathode (cell types 1, 2), or as an anode (cell type 3), or as both cathode and anode (cell type 4). The oxidizing property of conjugated polymers makes them suitable material as cathodes and their reducing property makes them suitable material as anodes in rechargeable batteries. As for example, the oxidizable conjugated polymers polypyrrole or polyaniline are used as a polymer cathode with a metal anode in the development of rechargeable (Figure 7.5, Type 1). In conjugated polymer batteries the charging and discharging process is nothing but the doping and dedoping process of the conjugated polymer. As for example [24], the charging

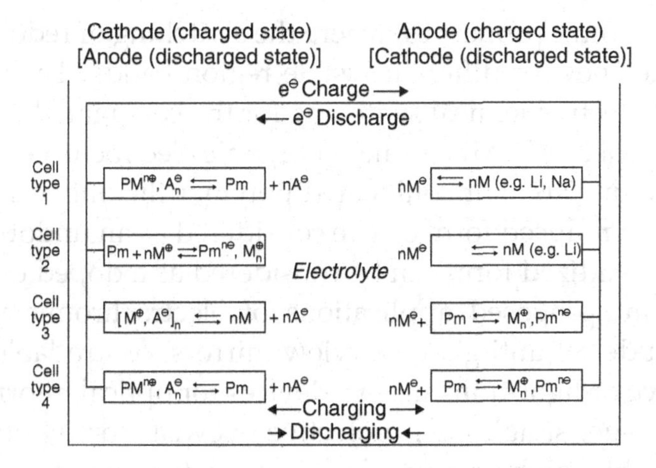

Figure 7.5 Different types of conjugated polymer battery with respective charging and discharging reactions. Reproduced with permission from ref. [90], Copyright © 1991 Elsevier Sequoia.

and discharging process can be explained as follows for a battery with lithium salt electrolyte, polyacetylene or polypyrrole as cathode and lithium as anode material.

Li	+LiY	+Conjugated polymer	$\underset{\text{Discharging}}{\overset{\text{Charging}}{\rightleftharpoons}}$	Li^{+} +	Conjugated polymer$^{+}Y^{-}$
(Anode)	(Electrolyte)	(Undoped; Cathode)		(Electrolyte)	(p-doped)

Similarly, the following charging and discharging reactions take place for a rechargeable battery consisting of Zn and polythiophene as the negative and positive electrodes respectively in ZnI_2 electrolyte.

Zn	+ZnI	+Polythiophene	$\underset{\text{Discharging}}{\overset{\text{Charging}}{\rightleftharpoons}}$	Zn^{2+}	+Conjugated polymer$^{+}I^{-}$
(Anode)	(Electrolyte)	(Undoped; Cathode)		(Electrolyte)	(p-doped by iodine)

7.6 Electrochromic Devices

Electrochromism is the reversible change of absorption spectra between two forms with distinct electronic (UV/visible) absorption spectra resulting from electrochemical (oxidation/reduction) reactions. Many organic materials exhibit electrochromic properties, where the switching of redox states generates new or different visible region bands. The mechanism of electrochemical switching for the conjugated polymer can be explained as the same for organic electrochromic materials (Figure 7.6). For conjugated polymer the only difference is that the reduced form can be considered as an undoped one and the oxidized form can be considered as a doped one. The important proposed applications of electrochromic materials include car anti-glare rearview mirrors, controllable light-reflective or light-transmissive devices for optical information and storage, sunglasses, protective eyewear for the military, controllable aircraft canopies, glare-reduction systems for offices, "smart windows" for use in cars and buildings, etc. All conjugated polymers are potentially electrochromic, i.e., redox switching, due to the transfer of electrons/counter anions giving rise to new optical absorption bands. In their intrinsic oxidized states, doped conjugated polymers in various doping levels possess a delocalized p-electron band structure [177]. The band gap, i.e., the energy gap between HOMO (valence band) and the LUMO (the conduction band), for conjugated polymers determines the intrinsic optical properties of these materials. Reduction of conducting polymers to an undoped

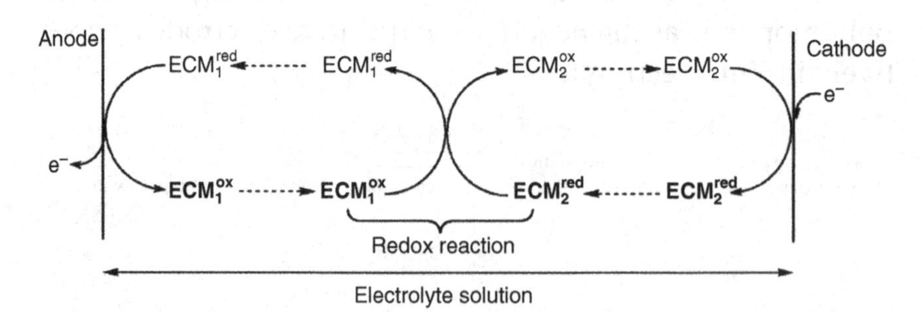

Figure 7.6 Operating principles of an electrochromic device (colored redox states of the electrochromic materials are shown in bold type; ECM: electrochromic materials; red and ox stands for reduced and oxidized state of the materials; for conjugated polymer oxidized state can be considered as doped and reduced state can be considered as undoped state. Reproduced with permission from ref. [177], Copyright © 1999 Elsevier Science Ltd.

form or lower intrinsic redox state removes the electronic conjugation which can affect the optical absorption band. The conjugated polymers can also undergo charge injected doping due to electron transfer to or from the electrode. The switching between neutral and oxidized states of bis(3,4-ethylenedioxythiophene)–(4,4′-dinonyl-2,2′-bithiazole) copolymer film shows electrochromic behavior due to the combination of an n-type electron-accepting monomer with a p-type electron-donating monomer [178]. Electrochromic properties of polyaniline have received greater attention due to their having a different intrinsic oxidation state according to doping level with distinct color. Polyaniline films are polyelectrochromic (transparent yellow-green-dark blue-black) involving protonation-deprotonation and/or anion ingress/egress giving the various redox states. In comparison with polyaniline, poly(o-toluidine) and poly(m-toluidine) films they have shown enhanced stability of polyelectrochromic response due to the lower conjugation length in those derivative [179]. Another good example is polypyrrole which is blue/violet (λmax~670 nm) in its doped (oxidized) state and yellow/green (λmax~420 nm) in undoped form [180]. Similarly, polythiophene thin films are blue (λmax~730 nm) in their doped (oxidized) state and red (λmax~470 nm) in their undoped form [181]. The electrochromic application of alkoxy-substituted polythiophenes is currently being investigated due to a band gap that is lower than polythiophene. The observation is that the conjugated polymers having phenyl ring with electron-withdrawing groups can stabilize the n-doped state. These conjugated polymers having phenyl ring with electron-withdrawing groups can also undergo a reversible reduction (n-doped) and oxidation (p-doped) process. As for example, a series of conjugated polymer films synthesized from 3-(p-α-phenyl)thiophene monomers (X=-CMe_3, -Me, -OMe, -H, -F, -Cl, -Br, -CF_3, -SO_2Me) have been studied for both the working and counter electrode materials in electrochromic devices [182].

7.7 Optoelectronic Devices

Currently, a wide band gap hole transport doped conjugated polymer is used in optoelectronic devices like organic light-emitting diodes (OLEDs) and organic solar cells (OSCs). The basic construction of devices in both cases is that an organic double layer is sandwiched between two electrodes acting as anode and cathode. The general operation principle of an OLED and an OSC is shown in Figure 7.7. According to the figure, the transport of charge carrier occurs to the active zone in the thin films for organic light-emitting diodes (OLEDs), while for organic solar cells (OSCs) it occurs away from the thin films. These systems are not operating by so called doping/dedoping phenomena, but a charge injection is occurring here. To make the charge transport efficient conductivity of the transport layers should be high and charge should be injected from either inorganic layer contacts (OLED) or withdrawn by such layer contacts (OSC). There are some important issues for improving the property of layers for these devices [183]:

a. The window layers of the organic semiconductor (conjugated polymer) can be used to optimize the optical properties with a maximum optical absorption field of the devices by optimizing the doping level.
b. In solar cells Ohmic contacts are more essential as any voltage drop considerably reduces the performance and doping can help to achieve better Ohmic contacts for the conjugated polymer.
c. Doping has an important role in efficient charge recombination between the layers.

Evidence of this is that poly(thienylene vinylene), poly(3,4-ethylene dioxythiophene)/poly(styrene sulfonate), (poly(3-hexylthiophene), etc., may be successfully introduced for the above devices by the use of electrically doped molecular

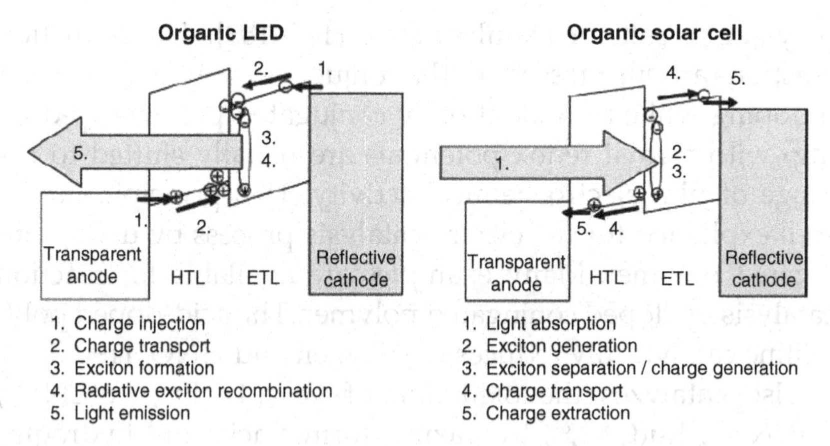

Figure 7.7 General operation principles of an OLED and an OSC. Reproduced with permission from ref. [183], Copyright © 2007 American Chemical Society.

charge injection layers. Poly(alkylbithiazoles) with unusual optical properties have received considerable attention for the application in LEDs because of their n-doping capabilities [184]. The improvement of photovoltaic response of the interface between poly(p-phenylene vinylene) and poly(ethylene dioxythiophene) has been reported with polystyrene sulfonic acid due to the effect of chemical doping of polymeric strong acid [185].

7.8 Others Applications

The conjugated polymers can be considered as membranes to separate gas or liquids due to their porosity. The oxidative doping causes the structural changes and an increase in the hydrophilicity of the conjugated polymer. For this reason an increase in water permeability properties was observed for doped polyaniline or polypyrrole films over the undoped (reduced) films, whereas poly(3-hexyl-thiopehe) shows a decrease in permeability for doping with increasing oxidation [186]. Corrosion protection for metal by doped conjugated polymer could be an important field of application. As for example, the passivation is achieved by acid-doped polyaniline salt for stainless steel [187–189]. The reduction reactions are suppressed for the conjugated polymers without n-doping while the oxidation of conjugated polymer (p-doping) with formal redox potentials are usually shifted to the range of film electrochemical activity. This principle can be well explained for the electro-catalysis process by using conjugated polymer. Many examples are available for reaction catalysis by doped conjugated polymer. The acid-doped polyaniline catalytically reduces the oxygen and HNO_3 [190, 191]. It also catalyzes the oxidation of Fe^{2+}, I^-, Br^-, $Fe(CN)_4^{6-}$, $W(CN)_4^{8-}$, $Ru(CN)_4^{6-}$, hydrazine, formic acid and hydroquinone. Catalytic reduction of oxygen and bromine on poly-(p-phenylene) is also reported [192, 193]. Exceptions to the simple relationship have been reported for iodine reduction on polythiophene or reduction of viologens on polyaniline which continue at more negative potentials, whereas the rates of ferrocene oxidation are even diminished on polythiophene in its conducting state [194–196]. The reason should be the generation of positively charged electronic species, i.e., hole injection at the polymer matrix by the reagents. Unexpected high hydrogen sorption for electrocatalysis at room temperature under 9.3MPa was observed on HCl-doped polyaniline and polypyrrole. According to the author, unusual hydrogen sorption results due to molecular sieving as well as stabilization of the conducting electronic environment [197]. The composites of conjugated polymers are preferably considered in electromagnetic interference applications (EMI), where the conjugated polymers can be used either as conducting filler for various insulating matrices or as an electrically conducting matrix with incorporated conducting/dielectric/magnetic materials. It is interesting to note that controlled doping leads to a marked improvement in dielectric properties over their undoped forms, although even after doping magnetic properties remain poor [198, 199]. A secondary mechanism of shielding is absorption by the electric or magnetic dipoles, which can interact with the transverse electric and magnetic vectors of the incident electromagnetic waves to introduce losses into the system. Doping of conjugated polymer leads to the formation of charged centers or improves the electric and magnetic vectors. Therefore, improvement of dielectric properties with doping level can be observed [198, 199]. Interestingly, doping-induced polarization or filler-induced interfacial polarization may contribute towards the dielectric properties of the conjugated polymer composites. For example, when conducting fillers like metal particles, graphite or carbon nanotubes are introduced into conjugated polymer matrices, further improvement of dielectric properties was observed due to induced interfacial polarization [198, 199].

8

Recent and Future Trends of Doping in Conjugated Polymer

8.1 Introduction

The advancement of science and technology provides smaller and smaller dimensions of materials with higher precision and better performance. Because of the requirement of smaller components for microelectronics, the conjugated polymers have captured interest as nanoscale objects. Thus, the recent development of doping in the conjugated polymer field has also been directed towards nanoscience and nanotechnology. Nanomaterials are materials with at least one dimension that is a nanometer in size, or more precisely, less than 100 nm. The transition from micro- to nanomaterials yields dramatic changes in various properties due to having a large surface area for a given volume, i.e., high aspect ratio. This is because many important chemical and physical interactions are governed by surface area and surface properties. Common geometries of nanomaterials and their respective

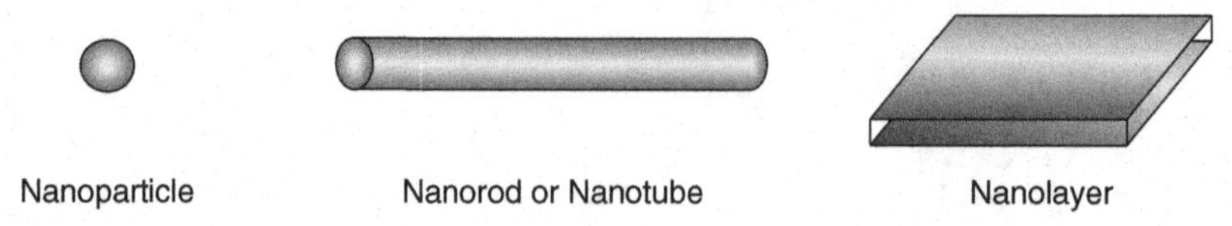

Figure 8.1 Types of common nanomaterials.

surface area-to-volume ratios are shown in Figure 8.1. In general, these materials are classified by their geometries; in broad terms the three classes are particle, layered, and fibrous materials. The fibrous materials may have structure like a wire (rod) or like a tube. Carbon black, silica, metallic nanoparticles and polyhedral oligomeric sislesquioxanes can be classified as nanoparticle reinforcing agents, while nanofibers and carbon nanotubes are examples of fibrous materials. When the filler has a nanometer thickness and a high aspect ratio (30–1000) plate-like structure, it is classified as a layered nanomaterial (such as an organosilicate).

Nanocomposites are those in which some nanomaterials are mixed inside a continuous matrix like polymer. Many investigations regarding the development of the incorporation techniques of nanoparticles into the polymeric matrices have been made. The properties of a nanocomposite are greatly influenced by the size scale of its component phases and the degree of mixing between the two phases. The above types of nanomaterials are successfully used to prepare nanocomposites with conjugated polymers. Depending on the nature of the components used (nanoparticle, nanolayered or nanofiber, and the conjugated polymer matrix) and the method of preparation, significant differences in composite properties may be obtained. The concept of nanotechnology can be introduced in the doping of conjugated polymer in two ways. The doping of nanostructured conjugated polymer, which may have better interaction with the dopant, can be employed or the conjugated polymer can be doped by the nanomaterials in the composite.

8.2 Doping of Nanostructured Conjugated Polymer

8.2.1 Introduction

One of the most promising ways to increase the performance of conjugated polymer is the nanostructured fabrication of the polymer, followed by subsequent doping. Nanostructured conjugated polymers have been subjected to recent investigation due to their unique combination of electroactive properties of conjugated polymers with the properties of nanomaterials. The nanoparticle of conjugated polymer is very rare as the polymeric big molecular chain preferably shows the fiber or tube geometry. Nano dimension, i.e., nanotube or nanofiber of conjugated polymers in doped form have been investigated due to their unique properties such as better conductivity and rapid reversible electrochemical switching. The doped nanostructured conjugated polymer shows high efficiency, even at relatively low doping ratios, due to amplified energy transfer caused by exciton diffusion. Another useful characteristic of nanostructured conjugated polymer is that a variety of dopants can be easily incorporated into the nanostructured polymer. Although there is no particular evidence, it is expected that the interaction between dopant and conjugated polymer will be enhanced in nanodimensions. Thus, the dopant can be helpful for the nanofabrication of the conjugated polymer. However, other problems like poor solubility, poor stability, poor conductivity, poor reproducibility, etc., remain as is. So, the nanofabrication of doped conjugated polymer can only have the better performances associated with the other problems in doped form.

8.2.2 Role of Dopant in Synthesis of Nanostructured Conjugated Polymer

The dopant itself is used to fabricate the nanostructured conjugated polymer successfully. Types of dopant interaction with

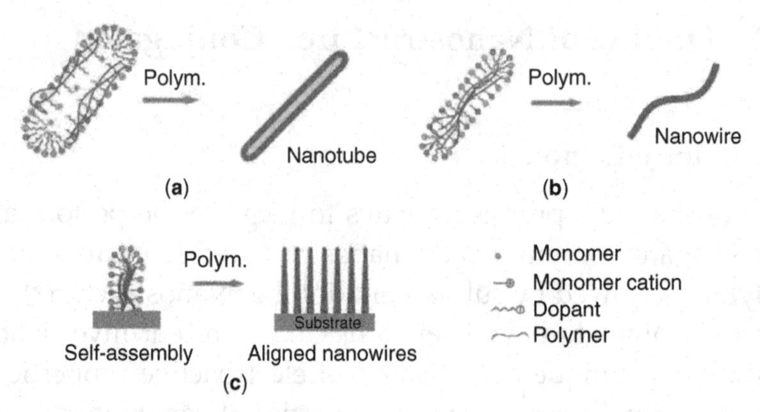

Figure 8.2 Types of dopant interaction with the conjugated polymer for the formation of (a) nanowire, (b) nanotube and (c) self-assembled nanowire. Reproduced with permission from ref. [200], Copyright © 2009 Elsevier Inc.

the conjugated polymer for the formation of nanowire, nanotube and self-assembled nanowire are shown in Figure 8.2. According to the figure it can be said that the mechanism of the formation of nanostructured conjugated polymer is influenced by dopant in three ways [200]:

a. Nanotubes of conjugated polymer are formed due to the synthesis of polymer on the surface of the self-assembled micelles of dopants and monomers.
b. Nanowires are formed due to polymerization of monomer inside the micelles surrounded by dopants.
c. Nanowires are formed due to synthesis of polymer carried out on the tips of the self-assembled layer of monomers and dopants

The nanostructured conjugated polymer can be synthesized on the surface of the various templates having nano-dimension channels, e.g., zeolites, nanoporous membranes, surfactants, polyelectrolytes or complex organic dopants. The hard templates like Zeolite channels, track-etched polycarbonate, anodized alumina, etc., are popular, whereas surfactants such as micelles, liquid crystals, etc., are well known as soft templates. The electrolyte, which can also serve as dopant for the synthesized conjugated polymer, can be used as soft-template for the synthesis of nanostructured conjugated polymer. The template of polyelectrolytes or complex organic dopants was successfully used to synthesize polyacetylene, poly(3-methylthiophene), polypyrrole and polyaniline nanofibers or nanotubes. The functionalized protonic acids D-10-camphorsulfonic acid, (4-{n-[4-(4-Nitrophenylazo)phenyloxy] alkyl}aminobenzene sul-fonic acid), azobenzenesulfonic acid, 5-aminonaph-thalene-2-sulfonic acid, etc., acts as dopant and templating agent for polyaniline nanotube synthesis. Similarly, poly(4-styrenesulphonate) has been used as a polymeric dopant as well as a template for polyaniline nanowire synthesis [201]. These nanoscale conjugated polymers can be successfully used for biosensors, electrochemical devices, single electron transistors, nanotips of field emission display, etc. The doped conjugated polymer can be directly synthesized using the hard dopant template with nano dimension. In core-shell nanoparticle synthesis the used template may serve as dopant counter ion for the synthesized conjugated polymer (Figure 8.3). As for example, the silica cores serve as templates for adsorption of aniline monomers as well as counter ions for doping of the synthesized polyaniline [202].

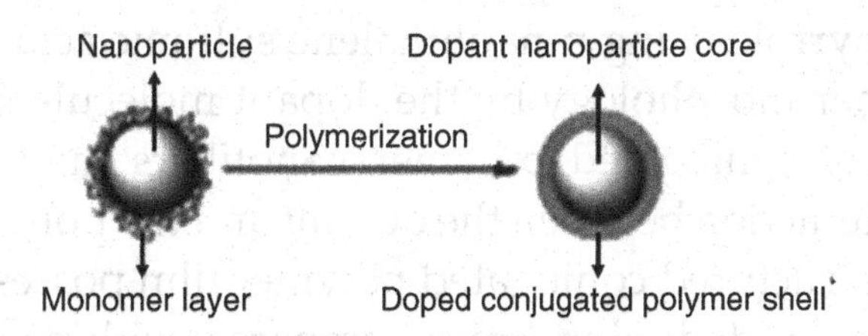

Figure 8.3 Nanoparticle shell of conjugated polymer doped with nanoparticle core.

Preformed polymers have been shown to self-assemble into nanoscale polymeric cylindrical structures due to molecular interactions such as hydrogen bonding and recognition with the dopant. For example, high molecular weight polyaniline has been changed to nanotubes by mixing with dopant camphor sulfonic acid and solvent 4-hexylresorcinol; self-organization into the polymer is induced due to the hydrogen bonding between all three molecules. Thin nanostructured films of sulphate-doped polypyrrole were chemically synthesized by interfacial polymerization to develop chemical and biological sensors by inserting various functional groups to polypyrrole films [203].

8.2.3 Property of Nanostructured Doped Conjugated Polymer

As a result of such nano-fabrication, the doped polymer can be twisted in proper orientation and thus its structural as well as electrical properties can be modified. It has been reported that the diameter of nanofibers, which are to some extent coupled and disordered as a result of doping, is strongly influenced by the dopant used in the polymerization. As for example, polypyrrole nanofibers were chemically synthesized via a functional dopant introduction process using p-hydroxy-azobenzene sulfonic acid as a functional dopant [204]. Similarly, polyaniline nanofibers were synthesized chemically from aniline [205]. A template-free method has been developed to synthesize microtubes of polyaniline and polypyrrole using p-naphthalene sulfonic acid to control the polymer morphology by the dopant molecule. The small diameter of conjugated polymer nanofibers may affect the proper interaction between the dopant and the polymer chain. The nanostructured conjugated polymer film possesses much faster doping/dedoping times compared with conventional cast film due to better interaction with the dopant in nano dimension. For example, the camphor sulfonic acid doped polyaniline nanofibers were shown to have a much faster response to doping/dedoping than conventional undoped polyaniline film. However, the environmental stability of the dopant counter ion inside the conjugated polymer with very small diameter is reduced. The camphor sulfonic acid-doped polyaniline nanofibers blended with polyethylene oxide are electrically insulating with a diameter below 15 nm, as the small diameter may allow complete dedoping in air exposure [206]. The nanotubes and nanowires of polypyrrole, polyaniline and polythiophene in doped and dedoped form were electrochemically synthesized using Al_2O_3 nanoporous templates for possible application as nanotip emitters in field emission displays and polymer-based transistors [207]. The dedoping of nanosystems in template-dissolving solvents has been observed as they were transformed from a conducting state to a semiconducting one.

8.3 Doping in Conjugated Polymer Nanocomposite

8.3.1 Introduction

Nanostructured conjugated polymers have been subjected to recent investigation due to the unique combination of electroactive properties of conjugated polymers with the properties of nanomaterials. However, these nanostructured conjugated polymers suffer from basic problems such as poor solution processability, poor environmental stability, poor mechanical strength and poor conductivity as well, which restrict the applications. It seems that a nanostructured conjugated polymer is not sufficient to solve the problem. Composites of conjugated polymers with nanoparticles have attracted recent interest due to their synergistic and hybrid properties derived from a wide range of components. The resulting nanocomposites have found successful applications with better performance than the conjugated polymer itself in versatile fields viz., battery cathode microelectronics, nonlinear optics, sensors, etc. However, insolubility and poor dispersibility of the nanomaterials inside the polymer matrix are major problems for its restricted application. Detailed knowledge on the interactions of the nanomaterials with polymer matrices will be helpful in order to develop, explain and correlate the properties of these nanocomposites for various applications with better performance.

8.3.2 Doping Interaction in Conjugated Polymer Composite with Nanoparticles

Utilization of nanomaterials as effective filler to improve the properties, especially the conductivity or mechanical property, of these conjugated polymer matrices is highly dependent on the homogenous dispersion of nanomaterials inside the polymer matrix. Conjugated polymer composites with various nanomaterials are of particular interest as they combine the properties of two or more different materials with the possibility of novel mechanical, electronic or chemical behavior. Such nanocomposites not only bridge the world of nanomaterials with that of the macromolecules, but also solve some of the problems associated with the doping of conjugated polymers. Novel properties of nanocomposites can be derived from the successful combination of the characteristics of parent constituents into a single material. The properties of the conjugated polymer composites with nanomaterials are strongly dependent on the effective interactions of the nanomaterials with polymer matrices. This homogenous dispersion of the nanomaterials is based on the effective strong or weak interaction with the polymer matrix. These interactions are also helpful to enhance the properties of the conjugated polymer composites with nanomaterials compared with that of the pristine conjugated polymers. Conjugated polymer nanocomposite offers a convenient way for obtaining reproducible and constant characteristic thin film.

Composites of conjugated polymers with a wide range of nanomaterials have attracted recent interest. In the nanoparticles doped conjugated polymer composite, the nanoparticles correspond to infinite or very large diffusion coefficients for the counter ions. The non-interacting nanoparticle can interrupt the doping interaction for the doped conjugated polymer. As for example, the conductivity decreases considerably when nanoclay is added to dodecylbenzene sulfonic acid-doped polypyrrole due to the interruption of effective doping [208]. The carbon black used as core material in its nanocomposites with conjugated polymer not only serves as the inorganic core but also helps in doping, due to the acid groups present in it [209]. The specific ion present in the nanoparticles can act as dopant counter ions for the conjugated polymer in nanocomposite like V^{5+} ions in the V_2O_5-polyaniline system. In other cases, some amount of surface charge carried by the Fe_3O_4 nanoparticles is transferred to the polyaniline in their nanocomposite by which Fe_3O_4 nanoparticles act as extra dopant [210]. Conjugated polymer metal nanocomposites show remarkable electronic and optical properties, which include high conductivities, electrochromism, electroluminescence and chemosensitivity. This is because resulting nanocomposites have found successful applications with better performance than the doped conjugated polymer itself in versatile fields viz., battery cathode microelectronics, nonlinear optics, sensors, actuators, etc. As for example, polyaniline *in situ* doped with Cr, Fe, Mn, Co and Al metal oxalates was chemically synthesized in aqueous sulphuric acid medium. Among the five metal oxalate complexes investigated, Al, Mn and Co upon doping into polyaniline with sulfuric acid dopant improve the polymerization yield, conductivity and thermal stability of polymer material. The large molecular size of the oxalates perhaps improves the efficient electron transfer and the consequent charge transport mechanism and support for crystallinity [211].

8.3.3 Doping Interaction in Conjugated Polymer Composite with Carbon Nanofibers or Nanotubes

An easy and effective solution for the improvement of the mechanical strength and conductivity of conjugated polymer is employed by making the nanocomposite with carbon nanotube (CNT). Conjugated polymer nanocomposites with CNT are used in many applications like organic light-emitting diodes, photovoltaic cells, energy storage devices, sensors, etc., with better performance. Utilization of CNT as effective filler with conjugated polymer is restricted due to the insolubility and poor dispersibility of CNT in common organic solvent, water and inside the polymer matrix. Functionalization of CNT (like –COOH) could be the solution for achieving homogeneous dispersion of CNT in conjugated polymer matrix through effective strong or weak chemical interaction. The nanocomposites of polyaniline with CNT have received great interest due to their charge transfer doping as well as site selective interaction with the polyaniline chain. The polyaniline-CNT nanocomposites were synthesized *in situ* and doped with protonic acid or without using any protonic acid dopant. However, polyaniline doped with protonic acid has been used for the preparation of nanocomposite and therefore the role of CNT as dopant is not clear. Again, the amount of CNT used is much higher, even up to 20%, for the nanocomposites of polyaniline with CNT only (without any protonic acid) [212]. Here the increase of conductivity for polyaniline-CNT nanaocomposite is mainly due to an increase of CNT concentration. In a recent study, the doping effect of carboxylic acid group functionalized multi-walled carbon nanotube (cMWCNT) has been investigated using polyaniline (PANI) without any other protonic acid dopant [99]. Here, the cMWCNTs were shown to have

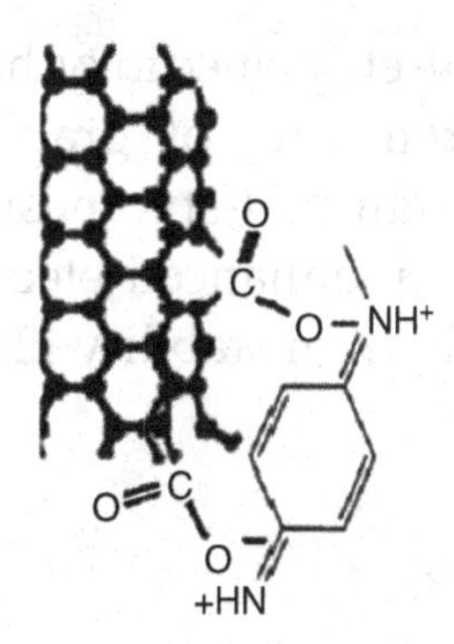

Scheme 8.1 Doping interaction between carboxylic acid functionalized multiwall carbon nanotube (cMWCNT) and polyaniline (PANI) chain. Reproduced with permission from ref. [99], Copyright © 2011 Elsevier B.V.

an effective dopant effect leading to an increase in conductivity of the nanocomposites (Scheme 8.1). Single-walled CNT/polyaniline composite films were prepared by electrochemical methods where donor–acceptor complex, i.e., weak redox doping results, improved electrochemical activity as well as conductivity of the composite [213]. Nanostructures of polyaniline composites containing metal oxide Fe_3O_4 nanoparticles were prepared with the help of napthalene sulfonic acid (NSA) dopant [214] for magnetic application. Similarly, polyaniline doped with different acids like H_3PO_4, H_2SO_4, $HClO_4$, CH_3COOH and CH_2=CHCOOH has been used to synthesize nanocomposite with Mn_3O_4 particles [215]. Polyaniline/inorganic nanocomposites, graphite/polypyrrole composites, polyaniline/activated carbon, p-doped 3-methylthiophene/activated carbon systems, graphene-polyaniline and ruthenium oxide/polypyrrole have been fabricated. The synthesis and characterization of graphene–polyethylene dioxythiophene nanocomposites have been investigated as electrode material for supercapacitor application [216]. The graphene–polyaniline composite material has also been synthesized using oxidative polymerization technique for supercapacitor application [217]. The composites of graphene oxide and poly(3,4-ethylenedioxythiophene) were prepared by *in-situ* polymerization of 3,4-ethylenedioxythiophene in aqueous mixture containing template of graphene oxide platelets modified with poly(sodium 4-styrenesulfonate). The resulting composites show an enhanced electrical conductivity of 9.2 S/cm with good thermal stability [218].

8.4 Future Trends

Though the doped conjugated polymer with low electrical conductivity has found immense use in the manufacture of semiconducting and dielectric devices, the aim of producing doped conjugated polymer with metallic conductivity has always engaged researchers. This is possible not only through the simple modification of organic conjugated polymers but also by combining them with efficient dopant. It is, therefore, no wonder that the doped conjugated polymers or synthetic metals with a wide range of products extending from most common consumer goods to highly specialized applications in space, aeronautics and electronics are being truly called the Materials of the 21st Century. The development of organic semiconductors as a new class of active materials for electronic and optoelectronic applications will take a few decades, just like the Si-based transistors, which are the basis of modern electronic devices, took about 50 years to develop. The joint efforts in materials development, fundamental research and device engineering will lead to the first commercial products using organic semiconducting materials.

The detailed study of doping in conjugated polymer offers many opportunities. The emerging field of conjugated polymers with special reference to their special features such as dopant type, doping process and dopant interaction, have been discussed briefly in this book. Although conjugated polymers have been synthesized and extensively studied in various laboratories around the world, they have not been commercialized yet. Both the stability and processability of doped conjugated polymers need to be substantially improved in order for their use in the marketplace to be possible. One of the fundamental challenges in the field of conjugated polymers is the design of doped conjugated polymer with high electrical conductivity or low band gap. The doped conjugated polymers with a rare combination of electrical, electronic, electrochemical and physical properties essentially demand an increase in their processability, environmental stability, thermal stability and good film forming property for real application purpose. The charge transfer characteristics of the conjugated polymers with various dopants are still not fully understood. Although extensive advances have been made during the last few years in the understanding of fundamental electronic, magnetic, spectroscopic, transport and theory relating to this extensive and diverse field of conjugated polymers, it is apparent that it has not yet been clearly explained. Thus, a better understanding of the relevant doping mechanisms is needed to support the drive towards the wide range of practical applications by solving these problems. The reproducibility of electronic, optical and photoelectrical properties, to mention a few, has made it possible to design and fabricate doped conjugated polymer-based devices. The richness of the chemistry, electrochemistry and physics of the conjugated polymer in doped form indicate they will continue to serve as a focus for challenging interdisciplinary research in the future.

The fields of conjugated polymers as well as nanotechnology are still in their infancy, although huge strides have been made over the last few years. Thus, the opportunities are endless for making exhilarating discoveries and inventing new technologies or devices based on these materials. Increasing interest in the practical use of nanostructured conjugated polymers and polymer nanocomposites has led researchers to make rapid developments in this field. It is proposed that the next stage of study should include coming up with devices that would have several advantages over the conventional material-made devices. The understanding of electrical transport in conjugated polymer nanofibers and nanotubes as well as conjugated polymer nanocomposite has improved, but still remains the subject of intensive discussion. Stability and processability both need to be improved for doped conjugated polymers, and the cost of these polymers should be reduced as well.

It can be predicted that the impact of conducting polymer materials on various applications will be dependent on the development of the ability to dope polymers that are repeatable, robust, and reliable in real environments. There is still the need for simple techniques that not only produce high-quality conjugated polymer nonofibers, nanotubes or nanocomposite, but also produce them in stable doped form. Thus, polymeric semiconductors, which are a must have in doped states are exciting materials and are essential in moving these new fields forward toward market applications. These novel doped conjugated polymers in macro or nano form having conducting and semiconducting properties will progress with time in becoming as commonplace as the insulating polymers are today.

References

1. V.R. Gowariker, N.V. Viswanathan, and J. Sreedhar, Polymer Science, 1st Ed.; New Age Publication (P) Ltd.: New Delhi 2005.
2. R.S. Potember, R.C. Hoffman, H.S. Hu, J.E. Cocchiaro, C.A. Viands, R.A. Murphy, and T.O. Poehler, *Polymers*, Vol. 28, p. 574, 1987.
3. M. Gauthier, M. Armand, and D. Muller, Electroresponsive Molecular and Polymeric Systems, Vol-1, Skotheim, T.A. Ed.; Marcel Dekker Inc.: New York, 1988.
4. F. Croce, S. Passerini, A. Selvaggi, and B. Scrosati, *Solid State Ionics*, Vol. 40-41, p. 375, 1990.
5. R.D. Armstrong, and M.D. Clarke, *Electrochem. Acta*, Vol. 29, p. 1443, 1984.
6. R. Dupon, B.L. Papke, M.A. Ratner, and D.F.Shriver, *J. Elecrtochem. Soc.*, Vol. 131, p. 586, 1984.
7. P.G. Pickup, *J. Chem. Soc. Faraday Trans.*, Vol. 86, p. 3631, 1990.
8. U. Salzner, J.B. Lagowski, P.G. Pickup, and R.A. Poirier, *Synth. Met.*, Vol. 96, p. 177, 1998.
9. Handbook of Conducting Polymers, Vol. 2, Skotheim, T.A. Ed., Marcel Dekker, New York 1986.
10. M. Gerard, A. Chaubey, and B.D. Malhotra, *Biosen. Bioelectron.*, Vol. 17, p. 345, 2002.
11. R.L. Greene, G.B. Street, and L.J. Sutude, *Phys. Rev. Lett.*, Vol. 34, p. 577, 1975.
12. H. Shirakawa, E.J. Louis, A.G. MacDiarmid, C.K. Chiang, and A.J. Heeger, *J. Chem. Soc. Chem. Commun.*, p. 578, 1977.
13. T.S. Moss, Handbook of Semiconductors, Vol. 1, Elsevier, Amsterdam 1992.
14. A.G. MacDiarmid, and A.J. Epstein, *Synth. Met.*, Vol. 69, p. 85, 1995.
15. K.S. Ho, *Synth. Met.*, Vol. 126, p. 151, 2002.
16. P.C. Wang, and A.G. MacDiarmid, *React. Fun. Polym.*, Vol. 68, p. 201, 2008.
17. R. Murugesan, and E. Subramanian, *Mater. Chem. Phys.*, Vol. 80, p. 731, 2003.

18. P.P. Sengupta, P. Kar, and B. Adhikari, *Thin Solid Films*, Vol. 517, p. 3770, 2009.
19. C.K. Chiang, A.J. Heeger, and A.G. MacDiarmid, *Ber. Bunsenges. Phys. Chem.*, Vol. 83, p. 407, 1979.
20. A.G. MacDiarmid, and A.J. Heeger, *Synth. Met.*, Vol. 1, p. 101, 1980.
21. M.M. Ayad, and E.A. Zaki, *J. Appl. Polym. Sci.*, Vol. 114, p. 1384, 2009.
22. M.V. Kulkarni, A.K. Viswanath, R. Marimuthu, and T. Seth, *Polym. Eng. Sci.*, Vol. 44, p. 1676, 2004.
23. A.G. MacDiarmid, R.J. Mammone, R.B. Kaner, S.J. Porter, R. Pethig, A.J. Heeger, and D.R. Rosseinsky, *Phil. Trans. R. Soc. Lond. A*, Vol. 314, p. 3, 1985.
24. D. Kumar, and R.C. Sharma, *Eur. Polym. J.*, Vol. 34, p. 1053, 1998.
25. A. Dall'Olio, G. Dascola, V. Varacco, and V. Bocchi *C R Acad. Sci. Ser. C*, Vol. 267, p. 433, 1968.
26. A.A. Athawale, M.V. Kulkarni, and V.V. Chabukswar, *Mater. Chem. Phys.*, Vol. 73, p. 106, 2002.
27. L.X. Wang, X.G. Li, and Y.L. Yang, *React. Func. Polym.*, Vol. 47, p. 125, 2001.
28. N.K. Guimard, N. Gomez, and C.E. Schmidt, *Prog. Polym. Sci.*, Vol. 32, p. 876, 2002.
29. G. Tourllion, and F. Garnie, *J. Phys. Chem.*, Vol. 87, p. 2289, 1883.
30. D.D. Borole, U.R. Kapadi, P.P. Kumbhar, and D.G. Hundiwale, *Mater. Lett.*, Vol. 57, p. 844, 2002.
31. D.D. Borole, U.R. Kapadi, P.P. Kumbhar, and D.G. Hundiwale, *Polym. Plast. Technol. Eng.*, Vol. 42, p. 415, 2003.
32. D.N. Debarnot, and F.P. Epaillard, *Anal. Chim. Acta*, Vol. 475, p. 1, 2003.
33. A.J. Dominis, G.M. Spinks, L.A.P. Kane-Maguire, and G.G. Wallace, *Synth. Met.*, Vol. 129, p. 165, 2002.
34. S. Kuwabata, K.I. Okamoto, O. Ikeda, and H. Yoneyama, *Synth. Met.*, Vol. 18, p. 101, 1987.
35. R. Mazeikiene, and A. Malinauskas, *Eur. Polym. J.* Vol. 36, p. 1347, 2000.
36. M.V. Kulkarni, A.K. Viswanath, R. Marimuthu, and T. Seth, *J. Polym. Sci. A: Polym. Chem.*, Vol. 42, p. 2043, 2004.
37. H. Hu, J.M. Saniger, and J.G. Banuelos, *Thin Solid Films*, Vol. 347, p. 241, 1999.
38. H.S. Moon, and J.K. Park, *Synth. Met.*, Vol. 92, p. 223, 1998.
39. Y.H. Liao, and K. Levon, *Polym. Adv. Technol.*, Vol. 6, p. 47, 1995.
40. S.A. Chen, and H.T. Lee, *Macromolecules*, Vol. 28, p. 2858, 1995.
41. Y. Cao, P. Smith, and A. Heeger, *Synth. Met.*, Vol. 32, p. 263, 1989.
42. H. Tsutsumi, S. Fukuzawa, M. Ishikawa, M. Morita, and Y. Matsuda, *J. Electrochem. Soc.*, Vol. 142, p. L168, 1995.
43. Y.H. Lee, J.Y. Lee, and D.S. Lee, *Synth. Met.*, Vol. 114, p. 347, 2000.

44. M.H. Lee, Y.T. Hong, and S.B. Rhee, *Synth. Met.*, Vol. 69, p. 515, 1995.
45. A.G. MacDiarmid, *Angew. Chem. Int. Ed.*, Vol. 40, p. 2581, 2001.
46. C.C. Han, and R.L. Elsenbaumer, *Synth. Met.* Vol. 30, p. 123, 1989.
47. R.C.D. Peres, J.M. Pernaut, and M.A.D. Paoli, *J. Polym. Sci., Polym. Chem.*, Vol. 29, p. 225, 1991.
48. S.M. Freund and A.B. Deore, Self-Doped Conducting Polymers, John Wiley & Sons Ltd, The Atrium, Southern Gate, Chichester, 2007.
49. A.O. Patil, Y. Ikenoue, F. Wudl, and A.J. Heeger, *J. Am. Chem. Soc.*, Vol. 109, p. 1858, 1987.
50. A.O. Patil, Y. Ikenoue, N. Basescu, N. Colaneri, J. Chen, F. Wudl, and A.J. Heeger, *Synthetic Metals*, Vol. 20, p. 151, 1987.
51. E.E. Havinga, L.W. Vanhorssen, W. Tenhoeve, H. Wynberg, and E.W. Meijer, *Polym. Bul.*, Vol. 18, p. 277, 1987.
52. Y. Ikenoue, Y. Saida, M. Kira, H. Tomozawa, H. Yashima, and M. Kobayashi, *Chem. Commun.*, p. 1694, 1990.
53. J. Yue, and A.J. Epstein, *J. Am. Chem. Soc.*, Vol. 112, p. 2800, 1990.
54. P. Kar, N.C. Pradhan, and B. Adhikari, *Synth. Met.*, Vol. 160, p. 1524, 2010.
55. P. Kar, N.C. Pradhan, and B. Adhikari, *Polym. Adv. Technol.*, Vol. 22, p. 1060, 2011.
56. J. Stejskal, M. Trchova, J. Kovarova, L. Brozova, and J. Prokes, *React. Func. Polym.*, Vol. 69, p. 86, 2009.
57. C.S. Priya, and G. Velraj, *Mater. Lett.*, Vol. 77, p. 29, 2012.
58. K.K. Kanazawa, A.F. Diaz, R.H. Gdss, W.D. Gill, J.F. Kwak, J.A. Logan, J.F. Rabolt, and G.B. Street, *J. Chem. Soc. Chem. Commun.*, p. 854, 1979.
59. K.K. Kanazawa, A.F. Diaz, W.D. Gill, P.M. Grant, G.B. Street, G.P. Gardini, and J.F. Kwak, *Synth. Met.*, Vol. 1, p. 329, 1980.
60. K. Gurunathan, A.V. Murugan, R. Marimuthu, U.P. Mulik, and D.P. Amalnerkar, *Mater. Chem. Phys.*, Vol. 61, p. 173, 1999.
61. W. Domagala, D. Palutkiewicz, D.C. Lacalle, A.L. Kanibolotsky, and P.J. Skabara, *Opt. Mater.*, Vol. 33, p. 1405, 2011.
62. C.J. Mathai, S. Saravanan, M.R. Anantharaman, S. Venkitachalam, and S. Jayalekshmi, *J. Phys. D: Appl. Phys.*, Vol. 35, p. 2206, 2002.
63. D.M. Hoffman, H.W. Gibson, A.J. Epstein, and D.B. Tanner, *Phys. Rev. B*, Vol. 27, p. 1454, 1983.
64. L.W. Shacklette, H. Eckhardt, R.R. Chance, G.G. Miller, D.M. Ivory, and R.H. Baughman, *J. Chem. Phys.*, Vol. 73, p. 4098, 1980.
65. K. Soga. Y. Kobayashi. S. Ikeda, and S. Kawakari, *J. Chem. Soc. Chem. Commun.*, p. 931, 1980.
66. M. Sato, K. Kaeriyama, and K. Someno, *Die Makromol. Chem.*, Vol. 184, p. 2241, 1983.
67. P. Kar, N.C. Pradhan, and B. Adhikari, *J. Polym. Mater.*, Vol. 25, p. 387, 2008.
68. Y. Kang, S.K.; Kim, and C. Lee, *Mater. Sci. Eng. C*, Vol. 24, p. 39, 2004.

69. K. Bienkowski, J.L. Oddou, O. Horner, I.K. Bajer, F. Genoud, J. Suwalski, and A. Pron, *Nukleonika*, Vol. 48, p. S3, 2003.
70. P.P. Sengupta, P. Kar, and B. Adhikari, *Bull. Mater. Sci.*, Vol. 34, p. 261, 2011.
71. Y. Shen, and M. Wan, *Synth. Met.*, Vol. 96, p. 127, 1998.
72. Y. Long, Z. Chen, N. Wang, Z. Zang, and M. Wan, *Physica B*, Vol. 325, p. 208, 2003.
73. L.H. Dao, M. Leclerc, J. Guay, and J.W. Chevalier, *Synth. Met.*, Vol. 29, p. 377, 1989.
74. C.G. Wu, Y.R. Yeh, J.Y. Chen, and Y.H. Chiou, *Polymer*, Vol. 42, p. 2877, 2001.
75. A. Kumar, R. Singh, S.P. Gopinathan, and A. Kumar, *Chem. Commun.*, Vol. 48, p. 4905, 2012.
76. N. Chanunpanich, A. Ulman, Y.M. Strzhemechny, S.A. Schwarz, J Dormicik, A. Janke, H.G. Braun, and T, Kratzmuller, *Polym. Int.*, Vol. 52, p. 172, 2003.
77. A. Sadek, W. Wlodarski, K.K. Zadeh, C. Baker, and R. Kaner, *Sens. Actuators A: Phys.*, vol. 139, p. 53, 2007.
78. A.Z. Sadek, C.O. Baker, D.A. Powell, W.Wlodarski, R.B. Kaner, and K.K. Zadeh, *IEEE Sens. J.*, Vol. 7, p. 213, 2007.
79. A.J. Heeger, *Synth. Met.*, Vol. 125, p. 23, 2002.
80. S. Hayashi, S. Takeda, K. Kaneto, K. Yoshino, and T. Matsuyama, *Jpn. J. Appl. Phys.*, Vol. 25, p. 1529, 1986.
81. W.M. de Azevedo, A.P. da Costa Lima, and E.S. de Araujo, *Radiat. Prot. Dosimetry*, Vol. 84, p. 77, 1999.
82. J.A. Malmonge, and L.H.C. Mattoso, *Synth. Met.*, Vol. 84, p. 779, 1997.
83. H. Bodugoz, and O. Guven, *Nucl. Instrum. Meth. Phys. Res. B*, Vol. 236, p. 153, 2005.
84. J.C. Scott, *J. Vac. Sci. Technol. A*, Vol. 21, p. 521, 2003.
85. D.B.A. Rep, A.F. Morpurgo, and T.M. Klapwijk, *Org. Electron.*, Vol. 4, p. 201, 2003.
86. B.H. Hamadani, H. Ding, Y. Gao, and D. Natelson, *Phys. Rev. B*, Vol. 72, p. 235302, 2005.
87. R.H. Friend, *Synth. Met.*, Vol. 51, p. 357, 1992.
88. A.C.R. Hogervorst, *Synth. Met.*, Vol. 62, p. 27, 1994.
89. S. Roth, H. Bleier, and W. Pukacki, *Faraday Discuss. Chem. Soc.*, Vol. 88, p. 223, 1989.
90. J. Heinze, *Synth. Met.*, Vol. 41-43, p. 2805, 1991.
91. P.P. Sengupta, and B. Adhikari, *Mater. Sci. Eng. A*, Vol. 459, p. 278, 2007.
92. P. Kar, A.K. Behera, N.C. Pradhan, and B. Adhikari, *Macromol. Sci. B: Phys.*, Vol. 50, p. 1822, 2011.
93. A.R. Hillman, S.J. Daisley, and S. Bruckenstein, *Electrochim. Acta*, Vol. 53, p. 3763, 2008.

94. K.J. Wynne, and G.B. Street, *Ind. Eng. Chem. Prod. Res. Dev.*, Vol. 21, p. 23, 1982.
95. M.D. Migahed, M. Ishra, T. Fahmy, and A. Barakat, *J. Phys. Chem. Solids*, Vol. 65, p. 1121, 2004.
96. B. Scrosati, *Prog. Solid State Chem.*, Vol. 18, p. 1, 1988.
97. D.W. Hatchett, M. Josowicz, and J. Janata, *J. Phys. Chem. B*, Vol. 103, p. 10992, 1999.
98. E.S. Blanca, I. Carrillo, M.I. Redondo, M.J. Gonzalez-Tejera, and M.V. Garcia, *Thin Solid Films*, Vol. 515, p. 5248, 2007.
99. A. Choudhury, P. Kar, *Composites: Part B*, Vol. 42, p. 1641, 2011.
100. L. Mihály, S. Pekker, and A. Jánossy, *Synth. Met.*, Vol. 1, p. 349, 1980.
101. J. Grobelny, J. Obrzut, and F.E. Karasz, *Synth. Met.*, Vol. 29, p. 97, 1989.
102. M. Sato, and H. Morji, *Macromolecules*, Vol. 24, p. 1196, 1991.
103. K. Mizoguchi, *Jpn. J. Appl. Phys.*, Vol. 34, p. 1, 1995.
104. X.L. Wei, M. Fahlman, and A.J. Epstein, *Macromolecules*, Vol. 32, p. 3114, 1999.
105. Y. Furukawa, *J. Phys. Chem.*, Vol. 100, p. 15644, 1996.
106. M. Kalbac, L. Kavana, and L. Dunsch, *Synth. Met.*, Vol. 159, p. 2245, 2009.
107. M. Kalbac, L. Kavana, M. Zukalova, and L. Dunsch, *Carbon*, Vol. 45, p. 1463, 2007.
108. P.G. Pickup, Electrochemistry of Electronically Conducting Polymer Films Modem Aspects of Electrochemistry, Number 33, edited by R.E. White et al. Kluwer Academic / Plenum Publishers, New York, 1999.
109. J.S. Miller, *Adv. Mater.*, Vol. 5, p. 671, 1993.
110. P. Novak, K. Muller, K.S.V. Santhanam, and O. Haas, *Chem. Rev.*, Vol. 97, p. 207, 1997.
111. W. Domagala, D. Palutkiewicz, D. Cortizo-Lacalle, A.L. Kanibolotsky, and P.J. Skabara, *Opt. Mater.*, Vol. 33, p. 1405, 2011.
112. B. Ulgut, J.E. Grose, Y. Kiya, D.C. Ralph, and H.D. Abruna, *Appl. Surf. Sci.*, Vol. 256, p. 1304, 2009.
113. O. Celikbilek, M. Icli-Ozkut, F. Algi, A.M. Onal, and A. Cihaner, *Organ. Electron.*, Vol. 13, p. 206, 2012.
114. B. Kaiser, *Adv. Mater.*, Vol. 13, p. 927, 2001.
115. J.P. Pouget, Z. Oblakowski, Y. Nogami, P.A. Albouy, M. Laridjani, E.J. Oh, Y. Min, A.G. MacDiarmid, J. Tsukamoto, T. Ishiguro, and A.J. Epstein, *Synth. Met.*, Vol. 65, p. 131, 1994.
116. P.F. vanHutten, and G. Hadziioannou, Crystallography of conductive polymers, in: H.S. Nalwa, (ed.) Organic Conductive Molecules and Polymers. Wiley, Chichester, UK, 1997.
117. K. Tashiro, M. Kobayashi, T. Kawai, and K. Yoshino, *Polymer*, Vol. 38, p. 2867, 1997.
118. J.P. Pouget, M.E. Jozefowiez, A.J. Epstein, X. Tang, and A.G. MacDiarmid, *Macromolecules*, Vol. 24, p. 779, 1991.

119. M. Salmon, A.F. Diaz, A.J. Logan, M. Krounbi, and J. Bargon, *Mol. Cryst. Liq. Cryst.*, Vol. 83, p. 265, 1982.
120. L.F. Warren, J.A. Walker, D.P. Anderson, C.G. Rhodes, and J. Buckley, *J. Electrochem. Soc.*, Vol. 136, p. 2286, 1989.
121. P. Saini, V. Choudhary, B.P. Singh, R.B. Mathur, and S.K. Dhawan, *Mater. Chem. Phys.*, Vol. 113, p. 919, 2009.
122. P. Saini, V. Choudhary, B.P. Singh, R.B. Mathur, and S.K. Dhawan, *Synth. Met.*, Vol. 161, p. 1522, 2011.
123. J.F. Rabolt, T.C. Clarke, K.K. Kanazawa, J.R. Reynolds, and G.B. Street, *J. Chem. Soc. Chem. Commun.*, p. 348, 1980.
124. N.T. Kemp, A.B. Kaiser, C.J. Liu, B. Chapman, A.M. Carr, H.J. Trodahl, R.G. Buckley, A.C. Partridge, J.Y. Lee, C.Y. Kim, A. Bartl, L. Dunsch, W.T. Smith, and J.S. Shapiro, *J. Polym. Sci., Part B: Polym. Phys.*, Vol. 37, p. 953, 1999.
125. M.H. Harun, E. Saion, A. Kassim, N. Yahya, and E. Mahmud, *JASA*, Vol. 2, p. 63, 2007.
126. J.A. Osaheni, and S.A. Jenekhe, *Chem. Mater.*, Vol. 7, p. 672, 1995.
127. N.E. Agbor, J.P. Creswell, M.C. Petty, and A.P. Monkman, *Sens. Actuators B: Chem.*, Vol. 41, p. 137, 1997.
128. C.K. Chiang, Y.W. Park, A.J. Heeger, H. Shirakawa, E.J. Louis, A.G. MacDiarmid, *J. Chem. Phys.*, Vol. 69, p. 5098, 1978.
129. A. Sakamoto, Y. Furukawa, and M.Tasumi, *J. Phys. Chem.*, Vol. 98, p. 4635, 1994.
130. M. Ando, C. Swart, E. Pringsheim, V.M. Mirsky, and O.S. Wolfbeis, *Solid State Ionics*, Vol. 152–153, p. 819, 2002.
131. E. Pringsheim, E. Terpetschnig, and O.S. Wolfbeis, *Anal. Chim. Acta*, Vol. 357, p. 247, 1997.
132. U.W. Grummt, A. Pron, M. Zagorska, and S. Lefrant, *Anal. Chim. Acta.*, Vol. 357, p. 253, 1997.
133. M.E. Nicho, M. Trejo, A.G. Valenzuela, J.M. Saniger, J. Palacios, and H. Hu, *Sens. Actuators B: Chem.*, Vol. 76, p. 18, 2001.
134. J.L. Bredas, J.C. Scott, K. Yakushi, and G.B. Street, *Phys. Rev. B*, Vol. 30, p. 1023, 1984.
135. L. Torsi, M. Pezzuto, P. Siciliano, R. Rella, L. Sabbatini, L. Valli, and P.G. Zambonin, *Sens. Actuators B: Chem.*, Vol. 48, p. 362, 1998.
136. S. Koul, R. Chandra, and S.K. Dhawan, *Sens. Actuators B: Chem.*, Vol. 75, p. 151, 2001.
137. N.M. Ratcliffe, *Anal. Chim. Acta*, Vol. 239, p. 257, 1990.
138. D.L. Ellis, M.R. Zakin, L.S. Bernstein, and M.F. Rubner, *Anal. Chem.*, Vol. 68, p. 817, 1996.
139. S. Virji, J. Huang, R.B. Kaner, and B.H. Weiller, *Nano Lett.*, Vol. 4, p. 491, 2004.
140. J.M. Slater, E.J. Watt, *Analyst*, Vol. 116, p. 1125, 1991.
141. Y. Cao, P. Smith, and A.J. Heeger, *Synth. Met.*, Vol. 48, p. 91, 1992.

142. R.H. Baughman, J.L. Bredas, R.R. Chance, R.L. Elsenbaumer, and L.W. Shacklette, *Chem. Rev.*, Vol. 82, p. 209, 1982.
143. M.K. Ram, O. Yavuz, and M. Aldissi, *Synth. Met.*, Vol. 151, p. 77, 2005.
144. W. Prissanaroon, L. Ruangchuay, A. Sirivat, and J. Schwank, *Synth. Met.*, Vol. 114, p. 65, 2000.
145. X.B. Yan, Z.J. Han, Y. Yang, and B.K.Tay, *Sens. Actuators B: Chem.*, Vol. 123, p. 107, 2007.
146. K.C. Persaud, *Mater. Today*, Vol. 8, p. 38, 2005.
147. L.B. Desmonts, D.N. Reinhoudt, and M.C. Calama, *Chem. Soc. Rev.*, Vol. 36, p. 993, 2007.
148. J.C. Chiang, and A.G. MacDiarmid, *Synth. Met.*, Vol. 13, p. 193, 1986.
149. D. Xie, Y. Jiang, W. Pan, D. Li, Z. Wu, and Y. Li, *Sens. Actuators B: Chem.*, Vol. 81, p. 158, 2002.
150. I. Kulszewicz-Bajer, A. Pron, J. Abramowicz, C. Jeandey, J.L. Oddou, and J.W. Sobczak, *Chem. Mater.*, Vol. 11, p. 552, 1999.
151. P. Kar, N.C. Pradhan, and B. Adhikari, *Sens. Actuators B: Chem.*, Vol. 140, p. 525, 2009.
152. P. Kar, N.C. Pradhan, and B. Adhikari, *J. Mater. Sci.*, Vol. 46, p. 2905, 2011.
153. H. Bai, and G. Shi, *Sensors*, Vol. 7, p. 267, 2007.
154. S. Watcharaphalakorn, L. Ruangchuay, D. Chotpattahanont, A. Sirivat, and J. Schwank, *Polym. Int.*, Vol. 54, p. 1126, 2005.
155. M.M. Verghese, K. Ramanathan, S.M. Ashraf, and B.D. Malhotra, *J. Appl. Polym. Sci.*, Vol. 70, p. 1447, 1998.
156. P.C. Pandey, and A.P. Mishra, *Analyst*, Vol. 113, p. 329, 1988.
157. M. Trojanowicz, W. Matuszewski, B. Szczepanczyk, and A. Lewenstam, in: G.G. Guilbault, and M. Mascini, Eds., Uses of Immobilized Biological Compounds. Kluwer, Dordrecht, 1993.
158. A.Q. Contractor, T.N. Sureshkumar, R. Narayanan, S. Sukeerthi, R. Lal, and R.S. Srinivasa, *Electrochim. Acta*, Vol. 39, p. 1321, 1994.
159. M. Nishizawa, T. Matsue, and I. Uchida, *Anal. Chem.*, Vol. 64, p. 2642, 1992.
160. K. Ramanathan, M.K. Ram, M.M. Verghese, and B.D. Malhotra, *J. Appl. Polym. Sci.*, Vol. 60, p. 2309, 1996.
161. K. Ramanathan, M.K. Ram, B.D. Malhotra, and A.S.N. Murthy, *Mater. Sci. Engg. C*, Vol. 3, p. 159, 1995.
162. K. Kaneto, M. Kaneko, Y. Min, and A.G. MacDiarmid, *Synth. Met.*, Vol. 71, p. 2211, 1995.
163. L. Kulinsky, H. Xu, H.K.A. Tsai, and M. Madou, *Proc. SPIE Int. Soc. Opt. Eng.*, p. 6173 (Smart Structures and Integrated Systems, 61730M/1-61730M/6) 2006.
164. T.F. Otero, and J.M. Sansinena, *Bioelectrochem. Bioenergy*, Vol. 42, p. 117, 1997.
165. T.F. Otero, and M.T. Cortes, *Sens. Actuators B: Chem.*, Vol. 96, p. 152, 2003.

166. M. Tahhan, V.T. Truong, G.M. Spinks, and G.G. Wallace, *Smart Mater. Struct.*, Vol. 12, p. 626, 2003.
167. G.M. Spinks, T.E. Campbell, and G.G. Wallace, *Smart Mater. Struct.*, Vol. 14, p. 406, 2005.
168. G.M. Spinks, B. Xi, V.T. Troung, and G.G. Wallace, *Synth. Met.*, Vol. 151, p. 85, 2005.
169. G.M. Spinks, V. Mottaghitalab, M. Bahrami-Samani, P.G. Whitten, and G.G. Wallace, *Adv. Mater.*, Vol. 18, p. 637, 2006.
170. V. Mottaghitalab, B. Xi, G.M. Spinks, and G.G. Wallace, *Synth. Met.*, Vol. 156, p. 796, 2006.
171. A. Facchetti, *Mater. Today*, Vol. 10, p. 29, 2007.
172. C. Bohn, S. Sadki, A.B. Brennan, and J.R. Reynolds, *J. Electrochem. Soc.*, Vol. 149, p. E281, 2002.
173. L. Torsi, M. Tanese, N. Cioffi, M. Gallazzi, L. Sabbatini, P. Zambonin, G. Raos, S. Meille, and M. Giangregorio, *J. Phys. Chem. B*, Vol. 107, p. 7589, 2003.
174. L. Torsi, A. Tafuri, N. Cioffi, M.C. Gallazzi, A. Sassella, L. Sabbatini, and P.G. Zambonin, *Sens. Actuators B: Chem.*, Vol. 93, p. 257, 2003.
175. L. Torsi, M.C. Tanese, N. Cioffi, M. Gallazzi, L. Sabbatini, and P.G. Zambonin, *Sens. Actuators B: Chem.*, Vol. 98, p. 204, 2004.
176. L. Torsi, A. Dodabalapur, L. Sabbatini, and P.G. Zambonin, *Sens. Actuators B: Chem.*, Vol. 67, p. 312, 2000.
177. R.J. Mortimer, *Electrochim. Acta*, Vol. 44, p. 2971, 1999.
178. F.C. Cebeci, E. Sezer, and A.S. Sarac, *Electrochim. Acta*, Vol. 52, p. 2158, 2007.
179. R.J. Mortimer, *J. Mater. Chem.*, Vol. 5, p. 969, 1995.
180. F. Garnier, G. Tourillon, M. Gazard, and J.C. Dubois, *J. Electroanal. Chem.*, Vol. 148, p. 299, 1983.
181. M. Mastragostino, in: B. Scrosati, Ed., Applications of Electroactive Polymers, Chapman and Hall, London, 1993.
182. D.J. Guerrero, X.M. Ren, and J.P. Ferraris, *Chem. Mater.*, Vol. 6, p. 1437, 1994.
183. K. Walzer, B. Maennig, M. Pfeiffer, and K. Leo, *Chem. Rev.*, Vol. 107, p. 1233, 2007.
184. S. Zhang, G. Nie, X. Han, J. Xu, M. Li, and T. Cai, *Electrochim. Acta*, Vol. 51, p. 5738, 2006.
185. A.C. Arias, M. Granstrom, D.S. Thomas, K. Petritsch, and R.H. Friend, *Phys. Rev. B*, Vol. 60, p. 1854, 1999.
186. I. Staasen, T. Sloboda, and G. Hambitzer, *Synth. Met.*, Vol. 71, p. 2243, 1995.
187. S. Shreepathi, H.V. Hoang, and R. Holze, *J. Electrochem. Soc.*, Vol. 154, p. C67, 2007.
188. Z. Deng, W.H. Smyrl, and H.S. White, *J. Electrochem. Soc.*, Vol. 136, p. 2152, 1989.

189. J.R. Santos, L.H.C. Mattoso, and A.J. Motheo, *Electrochim. Acta*, Vol. 43, p. 309, 1998.
190. G. Bidan, E.M. Genies, and M. Lapkowski, *J. Electroanal. Chem.*, Vol. 251, p. 297, 1988.
191. G. Mengoli, and M.M. Musiani, *J. Electroanal. Chem.*, Vol. 269, p. 99, 1989.
192. K. Ashley, D.B. Parry, J.M. Harris, S. Pons, D.N. Bennion, R. LaFollette, J. Jones, and E. King, *J. Electrochim. Acta*, Vol. 34, p. 599, 1989.
193. M.D. Levi, E.Y. Pisarevskaya, E.B. Molodkina, and A.I. Danilov, *J. Chem. Soc. Chem. Commun.*, p. 149, 1992.
194. M.D. Levi, and A.M. Skundin, *Sov. Electrochem.*, Vol. 25, p. 67, 1989.
195. A. Malinauskas, and R. Holze, *J. Electroanal. Chem.*, Vol. 461, p. 184, 1999.
196. G.C. Miceli, G. Beggiato, S. Daolio, P.G.D. Marco, S.S. Emmi, and G. Giro, *J. Appl. Electrochem.*, Vol. 17, p. 1111, 1987.
197. S.J. Cho, K. Choo, D.P. Kim, and J.W. Kim, *Catalysis Today*, Vol. 120, p. 336, 2007.
198. P. Saini, and V. Choudhary, *J. Nanopart. Res.*, Vol. 15, p. 1415, 2013.
199. P. Saini, and M. Arora, Microwave absorption and EMI shielding behavior of nanocomposites based on intrinsically conducting polymers, Graphene and Carbon Nanotubes in New Polymers for Special Applications Ed. by A.D.S. Gomes, inTech Publisher, Shanghai, 2012.
200. L. Xia, Z. Wei, and M. Wan, *J. Colloid. Interf. Sci.* Vol. 341, p. 1, 2010.
201. S. Jayanty, G.K. Prasad, B. Sreedhar, and T.P. Radhakrishnan, *Polymer*, Vol. 44, p. 7265, 2003.
202. J. Jang, J. Ha, and B.K. Lim, *Chem. Commun.*, p. 1622, 2006.
203. D.T. McQuade, A.E. Pullen, and T.M. Swager, *Chem. Rev.*, Vol. 100, p. 2537, 2000.
204. K. Huang, M.X. Wan, Y. Long, Z. Chen, and Y. Wei, *Synth. Met.*, Vol. 155, p. 495, 2005.
205. J. Huang, and R.B. Karner, *Angew. Chem. Int. Ed.*, Vol. 43, p. 5817, 2004.
206. Y. Zhou, M. Freitag, J. Hone, C. Staii, A.T. Johnson Jr., N.J. Pinto, and A.G. MacDiarmid, *Appl. Phys. Lett.*, Vol. 83, p. 3800, 2003.
207. B.H. Kim, D.H. Park, J. Joo, S.G. Yu, and S.H. Lee, *Synth. Met.*, Vol. 150, p. 279, 2005.
208. K. Boukerma, J.Y. Piquemal, M.M. Chehimi, M. Mravcakova, M. Omastova, and P. Beaunier, *Polymer*, Vol. 47, p. 569, 2006.
209. W.A. Wampler, K. Rajeshwar, R.G. Pethe, R.C. Hyer, S.C. Sharma, *J. Mater. Res.*, Vol. 10, p. 1811, 1995.
210. V.D. Pokhodenko, V.A. Krylov, Y.I. Kurys, O.Y. Posudievsky, *Phys. Chem. Chem. Phys.*, Vol. 1, p. 905, 1999.
211. R. Murugesan, and E. Subramanian, *Bull. Mater. Sci.*, Vol. 25, p. 613, 2002.

212. T. Jeevananda, Siddaramaiah, N.H. Kim, S.B. Heo, and J.H. Lee, *Polym. Adv. Technol.*, Vol. 19, p. 1754, 2008.
213. J.E. Huang, X.H. Li, J.C. Xu, and H.L. Li, *Carbon*, Vol. 41, p. 2731, 2003.
214. Z. Zhang, and M. Wan, *Synth. Met.*, Vol. 132, p. 205, 2003.
215. M.L. Singla, S. Awasthi, A. Srivastava, and D.V.S. Jain, *Sens. Actuators A: Phys.*, Vol. 136, p. 604, 2007.
216. F. Alvi, M.K. Ram, P.A. Basnayaka, E. Stefanakos, Y. Goswami, and A. Kumar, *Electrochim. Acta*, Vol. 56, p. 9406, 2011.
217. H. Gomez, M.K. Ram, F. Alvi, P. Villalba, E. Stefanakos, and A. Kumar, *J. Power Sources*, Vol. 196, p. 4102, 2011.
218. L.K.H. Trang, T.T. Tung, T.Y. Kim, W.S. Yang, H. Kim, and K.S. Suh, *Polym. Int.*, Vol. 61, p. 93, 2012.

Index

Also of Interest

Check out these published and forthcoming related titles from Scrivener Publishing

Atmospheric Pressure Plasma Treatment of Polymers
Edited by Michael Thomas and K.L. Mittal
Published 2013. ISBN 978-1-118-59621-0

Polymers for Energy Storage and Conversion
Edited by Vikas Mittal
Published 2013. ISBN 978-1-118-34454-5

Encapsulation Nanotechnologies
Edited by Vikas Mittal
Published 2013. ISBN 978-1-118-34455-2

Plastics Additives and Testing
By Muralisrinivasan Subramanian
Published 2013. ISBN 978-1-118-11890-0

Atmospheric Pressure Plasma for Surface Modification
By Rory A. Wolf
Published 2012. ISBN 9781118016237

Polymeric Sensors and Actuators
by Johannes Karl Fink
Published 2012. ISBN 978-1-118-41408-8

Handbook of Troubleshooting Plastics Processes
A Practical Guide
Edited by John R. Wagner, Jr
Published 2012. ISBN 978-0-470-63922-1

Introduction to Industrial Polypropylene: Properties, Catalysts, Processes
by Dennis P. Malpass and Elliot Band.
Published 2012. ISBN 978-1-118-06276-0

Antioxidant Polymers: Synthesis, Properties and Applications
Edited by Giuseppe Cirillo and Francesca Iemma
Published 2012. ISBN 978-1-118-20854-0

Handbook of Bioplastics and Biocomposites Engineering Applications
Edited by Srikanth Pilla
Published 2011. ISBN 978-0-470-62607-8

Biopolymers
Biomedical and Environmental Applications
Edited by Susheel Kalia and Luc Avérous
Published 2011. ISBN 978-0-470-63923-8

Renewable Polymers: Synthesis, Processing, and Technology
Edited by Vikas Mittal
Published 2011. ISBN 978-0-470-93877-5

Plastics Sustainability
Towards a Peaceful Coexistence between Bio-based and Fossil fuel-based Plastics
Michael Tolinski
Published 2011. ISBN 978-0-470-93878-2

Green Chemistry for Environmental Remediation
Edited by Rashmi Sanghi and Vandana Singh
Published 2011 ISBN 978-0-470-94308-3

High Performance Polymers and Engineering Plastics
Edited by Vikas Mittal
Published 2011. ISBN 978-1-1180-1669-5

Handbook of Engineering and Specialty Thermoplastics
Part 1: Polyolefins and Styrenics by Johannes Karl Fink
Published 2010. ISBN 978-0-470-62483-5
Part 2: Water Soluble Polymers by Johannes Karl Fink
Published 2011. ISBN 978-1-118-06275-3

Part 3: Polyethers and Polyesters edited by Sabu Thomas and Visakh P.M.
Published 2011. ISBN 978-0-470-63926-9
Part 4: Nylons edited by Sabu Thomas and Visakh P.M.
Published 2011. ISBN 978-0-470-63925-2

The Basics of Troubleshooting in Plastics Processing
By Muralisrinivasan Subramanian
Published 2011. ISBN 978-0-470-62606-1

A Concise Introduction to Additives for Thermoplastic Polymers
by Johannes Karl Fink.
Published 2010. ISBN 978-0-470-60955-2

Introduction to Industrial Polyethylene: Properties, Catalysts, Processes by Dennis P. Malpass.
Published 2010. ISBN 978-0-470-62598-9

Polymer Nanotube Nanocomposites: Synthesis, Properties, and Applications
Edited by Vikas Mittal.
Published 2010. ISBN 978-0-470-62592-7

Rubber as a Construction Material for Corrosion Protection
A Comprehensive Guide for Process Equipment Designers
2010. ISBN 978-0-470-62594-1

Miniemulsion Polymerization Technology edited by Vikas Mittal
Published 2010. ISBN 978-0-470-62596-5